Landscape Guide for Canadian Homes

Canada Mortgage and Housing Corporation

2004

CMHC offers a wide range of housing-related information. For details, call 1 800 668-2642
or visit our website at: www.cmhc.ca

Cette publication est aussi disponible en français sous le titre : L'amenagement paysager chez soi - Guide canadien, 63524

National Library of Canada cataloguing in publication data

Lefebvre, Daniel

Landscape Guide for Canadian Homes

This guide was prepared by Daniel Lefebvre and Susan Fisher.-Acknowledgements.
Issued also in French under title: L'aménagement paysager chez soi - guide canadien.
Includes bibliographical references.
ISBN 0-660-19278-0
Cat. no. NH15-422/2004E

1. Landscape design - Canada - Handbooks, manuals, etc.
2. Landscape gardening - Canada - Handbooks, manuals, etc.
3. Gardens - Canada - Design - Handbooks, manuals, etc.
I. Fisher, Susan, 1964- .
II. Canada Mortgage and Housing Corporation.
III. Title.

SB472.45L43 2004 712.6'0971 C2004-980097-3

Reprinted: 2005

Printed in Canada
Produced by CMHC

Acknowledgements

This guide was prepared by:
Daniel Lefebvre, Landscape Architect, Rousseau Lefebvre Architecture de paysage
Susan Fisher, Landscape Architect, Senior Researcher, CMHC

CMHC would like to acknowledge the following experts for their review and contribution of content to this guide:

Expert review of first draft:

- Christian Desmarais, Teacher and Horticulturist
- Daniel Glenn, Landscape Architect
- Stacy Moriarty, Landscape Architect
- Cecelia Paine, Landscape Architect

Expert review and content contribution on specific chapters:

- Jean-Marc Daigle, Landscape Architect
- Micheal D'Andrea, Water and Wastewater Services, City of Toronto
- Denis Flanagan, Landscape Ontario
- Tony Fleischmann, Certified Arborist
- Donna Havinga, Landscape and Environmental Consultant
- Henry Jun, Land Use Policy/Water Policy Branch, Ontario Ministry of Environment
- Peter Kreuk, Landscape Architect
- Cornelia Hahn Oberlander, Landscape Architect
- Peggy Lepper, Canadian Wood Council
- David Reid, Landscape Architect
- Stephane Savard, Pest Management Regulatory Agency
- Michael Simons, Engineer

CMHC would like to thank the following homeowners and others who contributed content for the testimonials:
Diane Brown (Delta Recycling Society), Tim Cross, Marjorie Mason Hogue, Jean Lamontagne, Friedrich Oehmichen, Jorg Ostrowski, Randy Penner, Doug Pollard, Frank Skelton, Gerald Sutton and Debra Wright.

The following CMHC researchers also contributed expert advice:
Sandra Baynes, Fanis Grammenos, Don Fugler, Thomas Green, Tom Parker, Ken Ruest, Virgina Salares, Darrel Smith and Ren Thomas.

Susan Fisher, Project Manager
Senior Researcher, Canada Mortgage and Housing Corporation

Table of Contents

Table of Contents

Table of Contents

MEETING YOUR NEEDS, WORKING WITH NATURE

Chapter 1

The yard–your little patch of outdoors where you play, barbecue, soak up the sun while eating, reading, chatting or watching the neighbours go by. Each space and feature should be designed to meet your needs including access, privacy, views, outdoor activities, relaxation, dealing with stormwater, making you feel closer to nature, expressing your style, and much more.

What is Healthy Housing:
Healthy Housing™ is a CMHC vision that promotes the health of occupants while protecting the environment and preserving our natural resources. Both indoors and in the landscape, a Healthy House conserves resources, for example through water and energy-efficiency, and through the durability and efficient use of building materials. A Healthy House is also affordable and adaptable to its occupants' changing needs. More information on Healthy Housing is available at **www.cmhc.ca**

Whether you want a radical makeover, are starting from scratch, want to spruce up an old garden or are just looking for maintenance tips, this book will help you meet your home landscape needs while respecting the natural environment, and saving time and money. The book shows how to go from ideas to plan to real life, by describing the design process and providing many tips on materials and technical solutions. These tips relate to a wide range of landscape elements like plants, soil, ponds, slopes, stormwater and hard surfaces such as decks and patios. These tips can be used by do-it-yourselfers or even those of you that are hiring a professional to design, prepare technical documents or install the design.

Figure 1-1

This book provides a process for identifying your needs and fitting them into the conditions that already exist on your property and in the natural environment around you. Many Canadians are becoming increasingly aware of the benefits of customizing their landscape

Figure 1-2

designs to the soil, sunlight, space and other conditions on their properties as well as the natural ecosystems and climate of their region. Rather than fighting nature by regularly mowing, spraying, draining and paving it, working with nature can save time, money and protect the environment.

What Are the Benefits?

The book provides solutions that include the following benefits:

A design that meets your needs: Create areas to allow outdoor activities, privacy, access and circulation, views, personal style, shade and sunlight, foundation drainage, etc.

Cost and time savings:

- Preserving the trees, soil and other features that are already on your property, will save you the expense and energy of removal, as well as buying, installing and establishing them.
- Choosing plants that are suited to your climate, soil, sunlight, space and moisture conditions will save you maintenance costs and time. Adopting the planting options, like xeriscapes, woodland and wildflower gardens, and maintenance techniques described in this book will also help. Watering, mowing, fertilizing, pruning, applying pesticides and hiring maintenance companies can eat up a fair amount of money and time.
- The value of a home can be increased by a well-designed landscape. Healthy, mature trees in particular add to a property's value.
- Using durable materials will help you save on long-term costs. Reusing and recycling materials can also save you money.
- Using compost produced at home can save you money on soil improvements.
- Well placed trees can reduce home cooling and heating costs. They can also reduce snow accumulation on your driveway and walkways.

Less water consumption: Water-hungry landscape practices, like over-watering your lawn or garden, put a strain on water supplies and municipal treatment facilities. These practices lower water tables and reduce stream flows, which affect fish, other animals and their habitats. Treating and supplying a growing demand for water, particularly in the summer peak demand periods, costs more for the municipality and ultimately, the taxpayer. That's why many municipalities enforce lawn watering restrictions and encourage citizens to cut down on water use.

Less harm to water quality and habitat occurs when there are fewer pavements that prevent stormwater from soaking into the ground. With impermeable surfaces, stormwater runs off quickly into sewer systems and sometimes overwhelms them. Stormwater runoff picks up road salt and oil, as well as fertilizers, pesticides and other substance which often end up back in our rivers and lakes, causing negative impacts to water quality and habitat. Impermeable surfaces also stop water from getting to the roots of your trees and other plants. (Figure 1-3)

Figure 1-3: Your landscape design and maintenance decisions have effects beyond your backyard, like the habitat and water quality of this neighbourhood stream.

More trees:

- Trees make you feel sheltered, private, comfortable and closer to nature. (Figure 1-4)
- Urban communities with trees are cooler in the summer than communities without trees. This is because plants evaporate water, so the sun's energy used for this process does not get re-radiated as heat. Trees also provide shade, which makes you feel cooler in the summer, and they provide wind protection during winter.
- They clean the air by filtering dust and absorbing air pollutants.
- They reduce stormwater runoff.
- They provide habitat for birds and other wildlife.

Figure 1-4: Healthy trees have many benefits. They make your property and neighbourhood more attractive and cooler on hot summer days. They provide habitat, shade and make us feel closer to nature.

Less air pollution, energy use and noise is generated from gas and electric-powered mowers, blowers and trimmers when you have low maintenance plantings.

Less pesticide use occurs by having diverse, disease-resistant plantings and by adopting the maintenance techniques, like non-synthetic pest controls, described in Chapter 7.

Healthier soils: Analyse your soil and add only appropriate amendments, like compost. It's free and it reduces the public costs and environmental impacts of waste management. Having healthy, dense plants, particularly on slopes, helps reduce erosion and cuts down on weeding. When fertile, healthy soil is washed off your lot, you lose a valuable resource and there are negative impacts on the water courses where it ends up.

When vegetation is replaced by surfaces like pavements, cities can be several degrees hotter in summer than nearby rural areas.

Use of healthy, durable building materials that come from your region: Using durable materials means that old materials need to be replaced less often, which saves wastage in the landfill and uses less energy in manufacturing and transporting replacement materials. This is also the case with reused and recycled materials. Using materials that were harvested and manufactured in your region also reduces transportation energy and pollution.

Did you know that by composting, you can reduce household waste by 30%. It's a valuable resource you can use to improve your soil.

Universal accessibility: It is easy to use, attractive, comfortable, safe and facilitates use by anyone in your household or visiting your home, including children, elderly people and those with mobility limitations. Think about how your needs might change. You may intend to keep your house into your retirement and beyond or as your parents age.

More locally unique natural and heritage features: Preserve and select materials and plant species that come from your region. Invasive

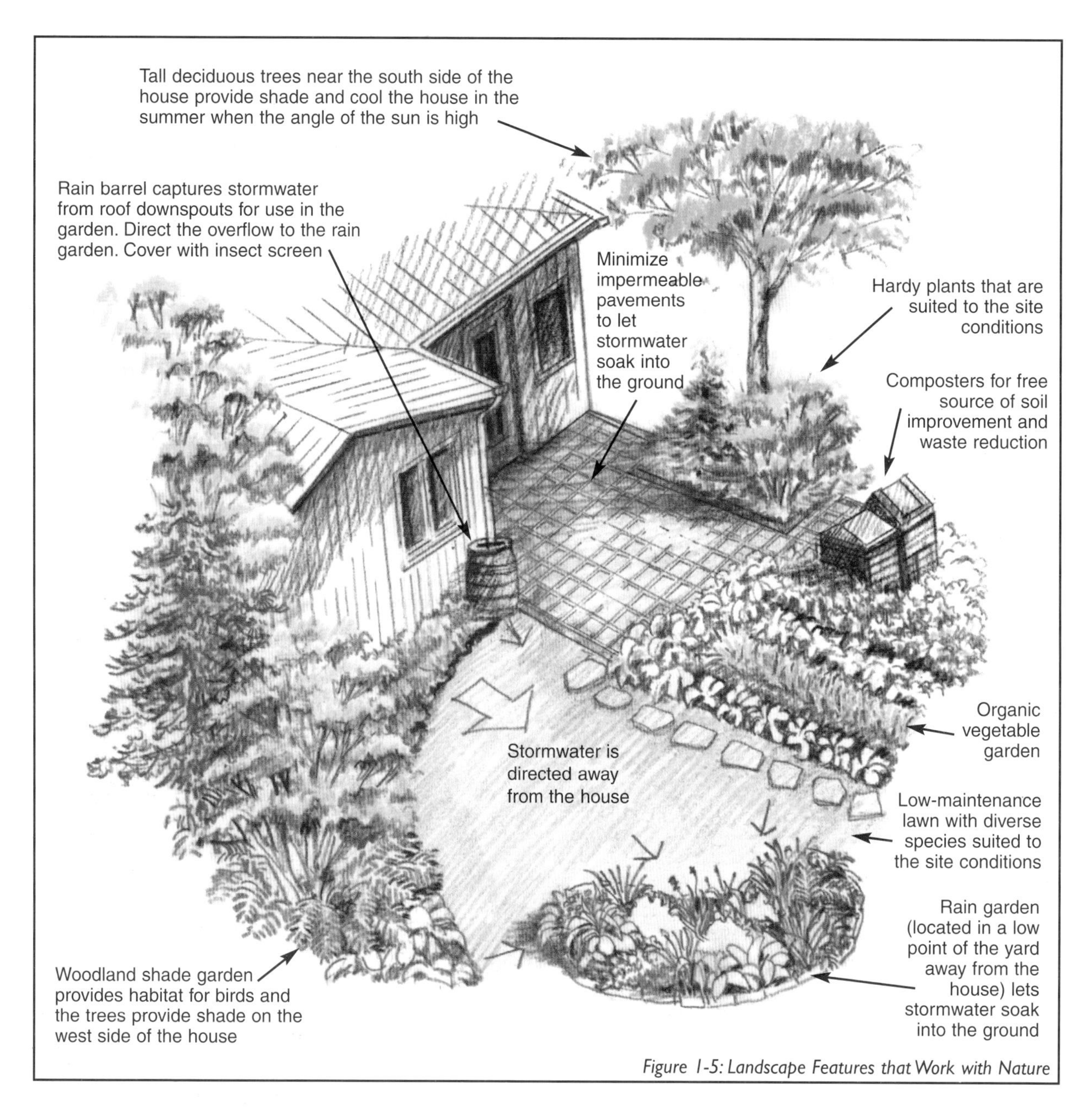

Figure 1-5: Landscape Features that Work with Nature

plant species can spread into natural areas, like a ravine or wood lot in your neighbourhood and can displace native species. This affects the diversity of the plant and animal life of your region.

Working With Nature

Consider a woodland evolving from a meadow. The infertile, exposed conditions gradually change as pioneer trees and shrubs grow and conditions become more fertile, protected and moist (Figure 1-6). From the trees in the upper canopy to the understorey species near the ground, the plants are adapted to the moisture, temperature, soil, and other environmental conditions in which they find themselves. Decaying plant materials, like fallen leaves, keep the soil moist and fertile, cycling nutrients back into the soil where they can be used by living plants. The trees provide a shady environment that is favorable to the plants underneath them and to their own seeds which can in turn give life to new trees. Trees around the edges protect other plants from exposure to wind and sun but provide just the right kind of environment for other plants to survive. Birds that find food and shelter in the forest help to spread seeds, and creatures living in the soil help to cycle nutrients. This landscape is adapted to natural

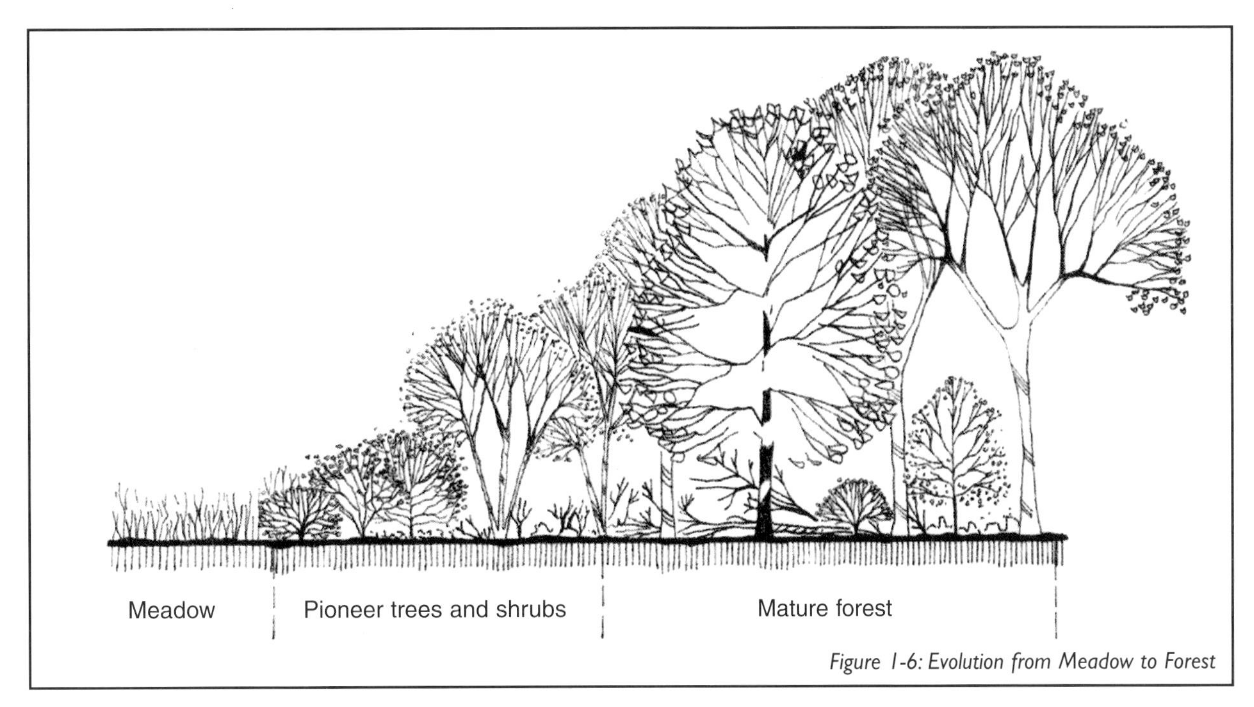

Figure 1-6: Evolution from Meadow to Forest

cycles and physical conditions, so nobody needs to water it, fertilize it, apply pesticides, mow, rake or trim it.

Many Canadians are establishing woodland gardens in their yards and benefiting from natural processes. (Figure 1-7) Similarly, many are inspired by the tall-, mixed- and short-grass prairies or wildflower meadows adapted to local rainfall, other weather patterns and soil conditions in certain parts of Canada. Even on a small backyard, these types of gardens can be very attractive, lively, dynamic, low-maintenance alternatives that harmonize with the natural environment.

Even if you are not recreating a woodland or a wildflower garden, you can take some clues from the plant communities that are adapted to your region. It could be as simple as choosing plants that are:

- suited to your local climate so that they will be healthier and generally require less maintenance, including water
- suited to your site in terms of soil conditions, sun and shade, high and low points, winds, space constraints, so that they will require less pruning, fertilizing, watering, etc.

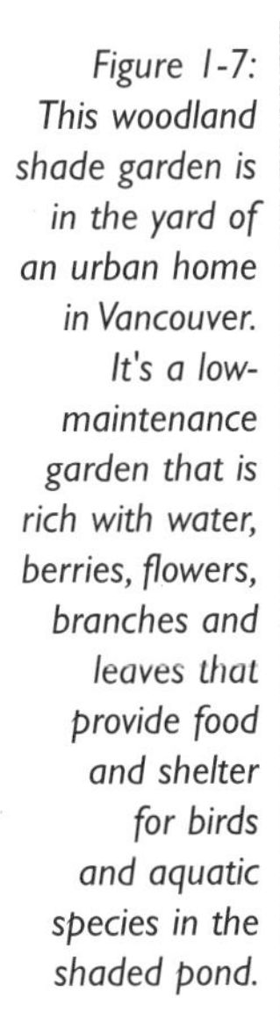

Figure 1-7: This woodland shade garden is in the yard of an urban home in Vancouver. It's a low-maintenance garden that is rich with water, berries, flowers, branches and leaves that provide food and shelter for birds and aquatic species in the shaded pond.

Photo: Frank Skelton

- diverse so they can better withstand pests. If there is an attack, only one species is affected and often, it can resist the attack if you choose pest-resistant species. Healthy plants are also more resistant to attacks.

In addition to choosing plants that are suited to their surroundings, taking clues from nature and letting it work for you can also involve:

- allowing decayed plant material to add nutrients, moderate soil temperatures and keep soil moist. Add organic mulch and keep chopped grass clippings and fallen leaves on the ground.
- minimizing surfaces that don't let stormwater soak in
- directing stormwater from your roof and elsewhere on your site to areas where water can soak into the ground
- preserving the soil on your property by analyzing it and selecting plants suited to those soil conditions, improving it as needed
- composting your kitchen waste and mulched garden waste, and using it to improve your soil (Figure 1-8)
- stabilizing slopes, particularly with plants, to reduce erosion

Figure 1-8: By composting, you can take advantage of a free source of organic matter for healthier soil and plants. Organic matter adds nutrients and improves soil structure, moisture and aeration.

- maintaining a healthy, dense plant cover and mulching around plants to help prevent weeds. Prevent pest problems by selecting the right plants. Identify pest problems correctly and get to know which insects and other critters are harmful and which are beneficial. For the harmful or undesirable ones, it helps to know when they emerge, what food they like and how to deal with them specifically in a way that won't harm anything else, for example by trapping pests or pulling weeds by hand.
- restoring habitat, like a woodland or creek bank or other shoreline on your property
- using building materials whose natural properties make them durable. Use local materials to reduce transportation energy and costs and to express your local identity. Also look for reused or recycled materials. Whenever you are replacing materials, like an asphalt driveway, find out if they are recyclable or reusable. Often this will save you tipping costs and keeps them out of the landfill. The manufacturing or harvesting processes of certain building materials also affect energy usage and the environment.
- fitting the design to your site by taking advantage of what's already there—trees, rocks, slopes, natural contours—will save you time and money, and provide great beauty and value to your property. Remember, a healthy, mature tree is irreplaceable.

Taking advantage of natural processes does not necessarily mean letting things go wild. It can, but this may not meet your needs or aesthetic standards—or those of your neighbours. This approach could involve a small lawn that has a mix of hardy species that are suited to the sunlight, water and soil conditions of your property. The way you maintain your lawn and garden is also important. It could be a water-efficient landscape, known as a Xeriscape™. (Figure 1-9). Or it could involve preserving a healthy tree that shades your home and attracts birds. It could mean having a driveway that

Figure 1-9: Xeriscapes, like this one in Peterborough, Ontario, are water-efficient landscapes that make use of a wide variety of low-water demand plants, mulch and efficient watering techniques. They not only save water but also maintenance time. Photo: Cathy Dueck, Peterborough Green-up

This is not only an environmental approach, but also an aesthetic and often an economical one. Replacing these features can be costly and difficult or impossible, particularly mature trees which take many years to grow. Also healthy plants already existing on your site have proven themselves to be suited to those conditions—climate, soil, water and sun. When plants are well suited to their site, you spend less time, energy and money maintaining them.

is as small as possible with a surface that lets stormwater soak in.

The term Xeriscape was developed by Denver Water in Colorado, U.S.A. to describe water efficient landscape practices. It is derived from the Greek word "xeros" meaning dry and "scape" derived from the word landscape.

There are infinite variations. Everyone has different needs and tastes—and your landscape should be customized to your needs. Every region of Canada has different plant communities, soils and weather patterns. So when it comes to landscapes that work with nature, it does not mean one design fits all. It simply means designing the landscape with the natural environment as a starting point.

Preserving What is Already There

For the new homebuyer or do-it-yourself homebuilder, it may be possible to plan changes to your future home if there are features you want to preserve before the home is built. Ask the following questions: Is it possible to keep a wooded area totally or partially intact? Can the house be located or designed to preserve existing trees, large rock outcroppings, natural contours and soils?

Considering the Natural and Cultural Context

Consider the features in your region or community when making your landscape choices, including historic or heritage features. Can your choice of new plants or materials be inspired by what you see in a nearby natural setting or in the community? Can soil, existing drainage and natural features be preserved or even emphasized? Consider ecosystems off of your property by creating links, no matter how small, with the local natural environment in order to help preserve it throughout the community.

By emulating the local natural environment in your yard, you can also provide habitat for birds, butterflies and other wildlife. This is both educational for children and puts us back in touch with nature. Landscape improvements to public spaces, like reforestation of a school yard or a park in your neighbourhood, can be done with the community's participation, and community members will truly feel a sense of ownership. Such a sense of belonging and responsibility facilitates a project's long-term success. This book will focus on individual residential properties, but many of the concepts can be applied to community landscapes as well. (Figure 1-10)

How Is This Book Organized?

The book is organized according to the order normally followed in the design and installation process. Chapter 2 describes the design process, followed by Chapter 3, which shows how to analyze and amend soil and provides advice on grading, slope stabilization and retaining walls. Chapter 4 deals with water, including stormwater management, water ponds, irrigation and shoreline habitat restoration. Chapter 5 provides advice on hard surfaces, like driveways, patios, decks and walkways. Chapter 6 provides tips on planting design and a section on tree preservation, followed by Chapter 7, which provides plant installation and maintenance advice. The final chapter offers tips for hiring contractors.

Figure 1-10: While this book focuses on individual homes, many of the concepts can be applied to community landscape projects, such as the reforestation of parks and schoolyards.

Photo: Evergreen

The Residential Yard as an Ecosystem

Picture your yard as an ecosystem. It is nested within a larger ecosystem—a local watershed, which is part of a regional watershed, a bioregion, a life zone, and ultimately, the biosphere. When pictured this way, it may be possible to see and appreciate the myriad connections that link the yard, and your actions within it, to the surrounding landscape. Rain that falls in the yard is typically conveyed off-site, over land or through sewers, to low-lying areas nearby, while small amounts may filter through the soil to replenish groundwater. Similarly, the air we breathe moves with the wind and thus transcends artificial boundaries, as do songbirds, butterflies, and bees darting from one garden to the next to nearby natural areas in search of food and shelter. These and a multitude of other ecological relationships are powerful reminders of the interconnectedness of all things and of the need to be responsible stewards in our interactions with the land.

DESIGNING YOUR LANDSCAPE—THE PROCESS

The design possibilities on a single site are numerous. Each landscape plan or design will be different, depending on your needs, your personal taste, your property conditions, your budget and legal requirements. The planning approach in this book has eight steps, aimed at finding the best fit among these factors. These steps will help you prepare your own design but they will also be helpful if you are hiring a design professional, such as a landscape architect or landscape designer, to prepare the design and technical documents and oversee the installation. These steps are useful whether you are doing the installation yourself or hiring a professional to do it. More details on hiring professionals are provided in Chapter 8.

The following chart summarizes the eight-step design process.

Following these steps will help you take advantage of the features you already have on your property and find the best fit with your needs. This will save you time and money and is better for the environment.

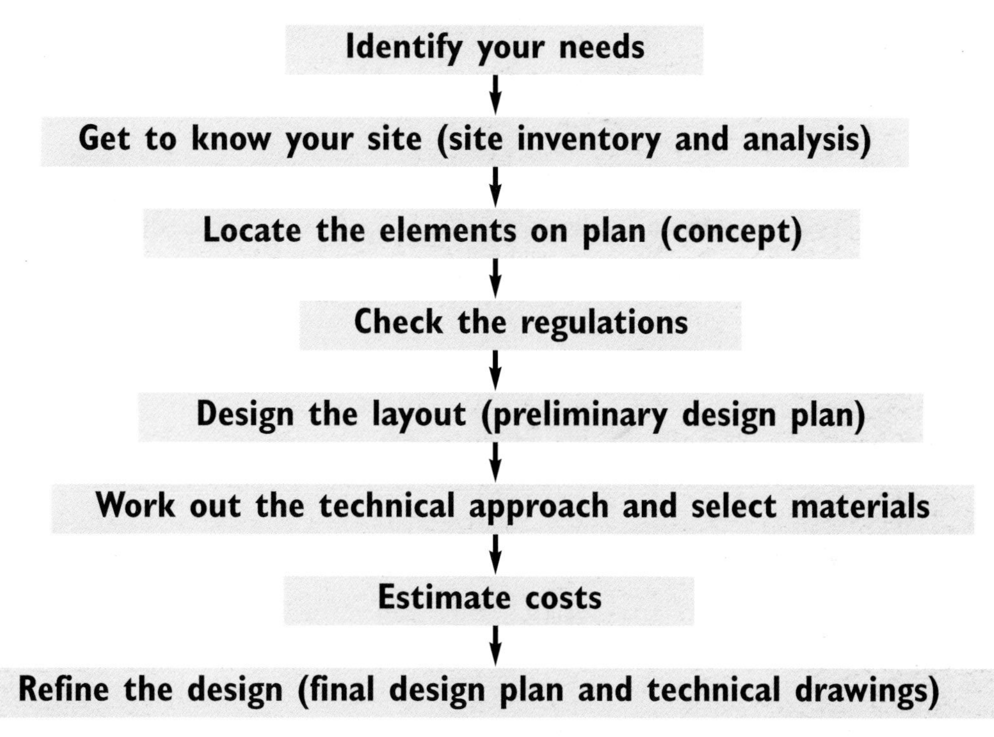

Step 1. Identify Your Needs

What are the activities and features you want to incorporate into your design, for example, an eating area or a vegetable garden? What are the space requirements and other qualities needed for these features? Here is a list of activities and features to consider in designing your landscape.

Access and circulation:

- pedestrian access from the sidewalk, street or lane to your home's entryways. Also consider wheelchair, bike and rollerblade access.
- foot travel within your yard, for example, from the back door to the backyard or from the patio to the vegetable garden. Also consider wheelchair access.

- vehicle access from the street or rear lane to the outdoor parking area or garage
- parking for your household and visitors. Account for the number and size of vehicles. It's best to keep this to the minimum required on a daily basis only.
- access prevention. Do you want to prevent access to certain areas?

For many years, public places have been designed or adapted to include access for people with mobility limitations. Universal design can apply to your private yard. Consider eliminating stairs, steep grades or surfaces that make circulation difficult for people with mobility limitations. You'll find that this generally makes access easier and safer for you and your family, for example when using bicycles and wheelbarrows.

For many Canadian homes, a large part of the front yard is taken over by the driveway. But through careful assessment, you can limit the size of your driveway to what is required for parking and garage access on a daily basis only. You can then maximize your yard space for other uses, such as play areas or plantings. Minimizing pavements like asphalt, also helps reduce stormwater runoff and allows water to soak into the ground, which in turn protects our natural waterways and puts less strain on our municipal water systems. Areas with a high proportion of plants, especially trees, also have lower temperatures in summer than areas with higher proportions of paved surfaces.

Outdoor activities:

How much space and what features would you like for the following areas:

- space allocated for eating and gathering, such as a patio, deck or barbecue area
- open areas for activities, like tossing a ball or playing with the dog
- active play areas with play equipment, like a climbing set or sand box
- quiet sitting areas
- vegetable and fruit production
- areas for attracting birds, butterflies and other wildlife
- special gardening interests, such as flowers, ponds or woodland gardens

Allowing space for these activities is an essential part of the process. Assessing your needs will help identify your space requirements, and where possible will help combine space for activities that are compatible. Ask yourself which activities you want to include, and what their relative importance will be, as your needs change.

Consider what qualities you would like these areas to have; for example a sunny or shaded location or an open or protected spot. Also think about what equipment or furniture you need, and where they should go in relation to other places. For example, you may want to locate your eating area with easy access to the kitchen. The chapters that follow provide tips, such as space requirements and best locations for a range of outdoor activity areas.

Aesthetic qualities and views:

- visual focal point. Are there areas or objects you want to draw attention to, for example, a colourful flowerbed, architectural feature, sculpture or special tree?
- views. Are there parts of your yard or surroundings you want to see from the house, or views you want to create from the garden or from the street to your house? Are there undesirable views you want to block?
- privacy. Are there places where you don't want to be seen from outside your lot?
- areas of sun, shade or semi-shade
- personal taste. Do you want open, low areas or intimate areas with a canopy overhead; straight lines or curves, simplicity or complexity?

Other elements:

- services and infrastructure, like lighting and water/irrigation
- storage space. Do you need an outdoor shed or shelves to store gardening tools and snow removal equipment? If so, what size?
- drainage and infiltration (areas where water should drain from, for example, away from the house, and areas where you want rainwater to soak in)

- garbage receptacle
- composting
- snow storage
- water storage, for example, rain barrels

Step 2. Get to Know your Site

A detailed analysis of your site will help you take full advantage of the features both on it and around it. For example, you can take advantage of a view to a neighbour's tree, even if it is not in your yard. The information gained from the analysis will help guide the design, for example by indicating the best place for a patio to offer the best views or the sunniest place for a vegetable garden. Knowing your site is also the first step to taking advantage of the features that already exist: vegetation, slopes, views, soils and more. Ultimately, this saves you time, money and is better for the environment.

What is scale: When drawing "to scale" every unit of measurement on paper, for example, a centimetre, represents a corresponding unit of measurement in reality, like a metre. In this case, the scale is 1:100.

Start by drawing a base plan of the site on which you can locate the different features of your property and its immediate surroundings. A good place to start is the legal survey that you may have obtained when you purchased your home. Surveyors prepare these plans using precise instruments and they are to scale. They note the location and dimensions of the lot, the buildings and the various easements. If you don't know the limits of your property, it's best to hire a surveyor to locate them before you begin the work. Placing a feature like a fence on your neighbour's property can cause you problems later.

You can measure your property's features and sketch them onto a plan yourself. The list below indicates the elements you should consider when preparing your inventory and site analysis:

Inventory and Analysis—Suggested information to collect:

Topography (see Chapter 3)
- indoor level of floor at doorways to the house
- high and low points of your property
- slopes
- main shapes and high/low point of special topographical features, like a berm
- level at the corners of your lot
- level at sidewalk or if none, at curb or street edge
- level at top and bottom of existing stairs, walkways, porches and decks

Soils (Chapter 3 provides a process for analyzing your soil)
- quality of existing soil, including fertility, texture, pH, structure, moisture, etc.
- location of any erosion
- underground geology, such as depth to bedrock
- special characteristics, like rock outcroppings

Drainage (see Chapter 4)
- natural drainage elements, like streams, ponds or wetlands
- drainage swales and ditches
- moist areas where water collects
- direction of surface runoff
- water table (depth, seasonal variation)

Vegetation (see Chapter 6 and 7)
- trees: location, species, condition and size, including the trunk diameter, height and span of the crown. Inventory shrubs as well.
- outline of areas of flowerbeds, perennial borders, lawn and vegetables gardens
- unique specimens, plant groupings
- special natural features, like woodlands

Microclimate
- areas of shade and sun
- direction of prevailing wind and wind-exposed areas
- quantity of snow and areas of snow accumulation
- local rainfall conditions

Existing structures
- buildings including the house (also locate

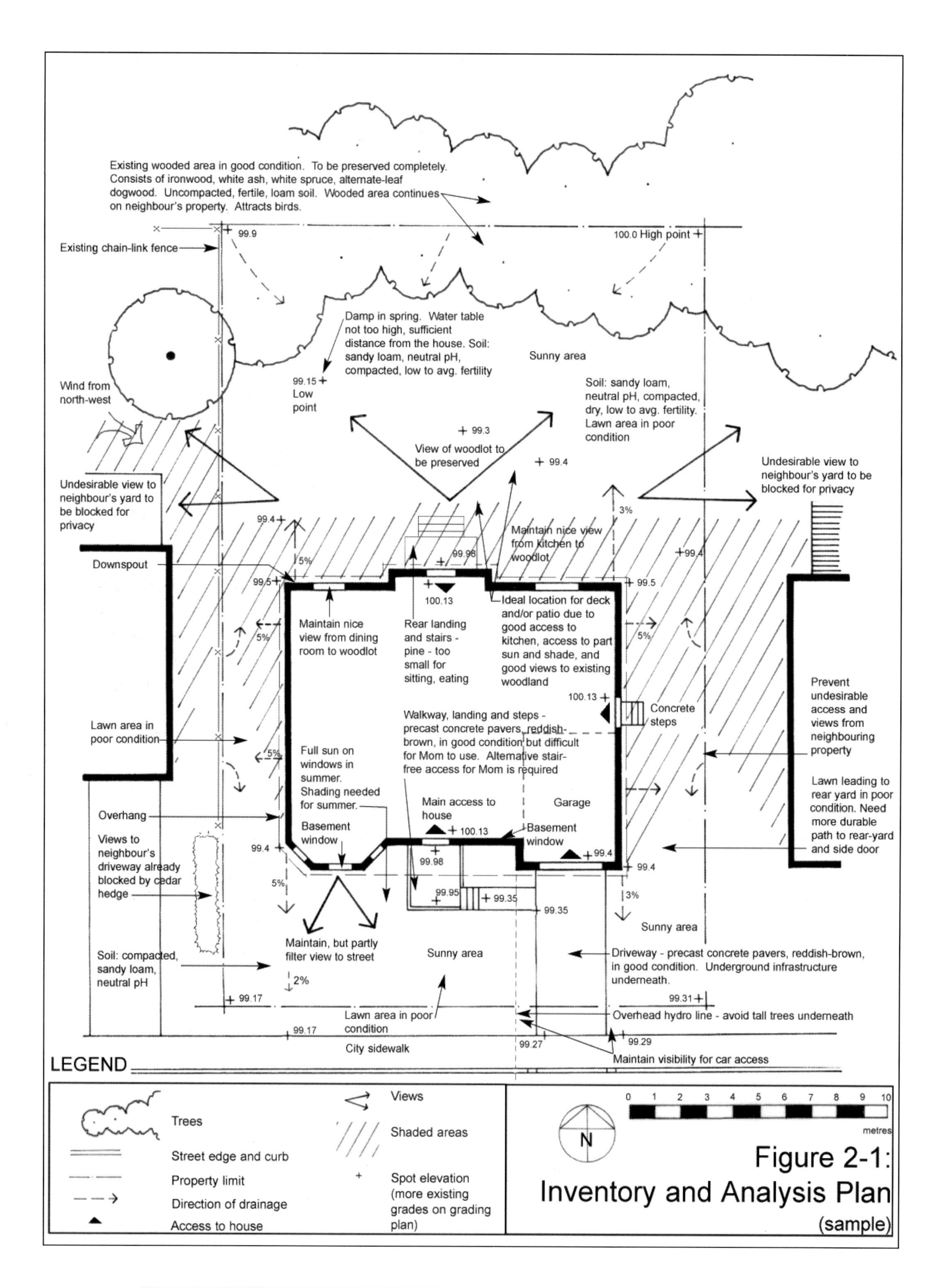

Figure 2-1: Inventory and Analysis Plan (sample)

exterior doors, windows, basement wells, downspouts, roof overhangs), shed and garage
- patios, porches, decks, and fences
- special architectural features

Access
- exterior doorways to the house and other buildings on the site
- paths within the property and from off the lot, like a city sidewalk
- driveway or laneway

Views
- views to neighbouring lots from your property, both desirable and undesirable
- views to your property from neighbouring lots, both desirable and undesirable

Elements of neighbouring properties
- use of the adjacent lands (type, impact on your site)
- location of woods and other natural features
- adjacent streets, lanes and sidewalks
- location of buildings and special features on adjacent lots

Other
- underground infrastructure (water, sewer, hydro, cable, phone), right-of-ways, easements

> Move around the garden at different times of day to identify the locations exposed to sun, shade or wind. The sun's position changes with the seasons. So where possible, conduct this examination at different times of the year. Design features like vegetation can enhance your comfort, for example by blocking winds and providing shade.

In addition to the location of these elements, it is also important to note their quality. The analysis involves qualifying or judging the features you have inventoried. It also indicates where you can make improvements, like views to open up or close, areas where you want more or less shade, or where there are drainage problems. These notes can be written on an inventory and analysis plan, after you have completed your inventory.

This information will prove to be an excellent guide when developing your concept. By using all of the information gathered, you will more easily be able to place your patio so that it offers a good view, or position your vegetable garden in a sunny area.

Figure 2-1 provides an example of an inventory and analysis plan.

Step 3. Locate the Elements in a Concept Plan

You will want to determine the best location and arrangement for specific activity areas, footpaths and other features. To get to this stage, a concept plan or "bubble" diagram is a useful tool (Figure 2-2). It involves placing the physical elements you identified in your needs assessment, like the driveway, patio, vegetable garden, etc. into a plan of your property. Each use or activity area is represented by a bubble placed in relation to the features you identified in your site inventory and analysis, including views and access points. Through this process, you can explore several options or scenarios and you pick the one that is the best fit between your site and your needs. **Use trace paper placed on top of your site inventory/ analysis plan or make several photocopies of it, so that you can try out different design options.**

The main elements to locate on your concept plan are:
- pedestrian movement patterns to the house and around the yard
- recreational activity areas
- areas for sitting/meals, like patios and decks
- main grading and drainage elements, areas for stormwater infiltration and storage
- view screening (such as fences and hedges) and view enhancement
- climate protection (wind, sun, rain)
- links with surrounding ecosystems
- focal points
- special features, like a water garden, flower bed or wooded area
- vegetable gardens and composters
- access barriers, like fences
- vehicle access, circulation and parking
- storage areas

The relationships between activities and existing features should be resolved at this conceptual stage in order to reduce the number of incompatible conditions. Once you find the best solution, the concept plan will be the basis for a more detailed and refined plan.

Step 4. Check the Regulations

To avoid future problems it is essential to consult the regulations near the beginning of the landscape planning process.

Municipal regulations

Municipalities are the main agency that regulates landscape features you may be including in your design. Generally, municipal offices are the best resource for information on the regulations applicable to residential landscapes.

In most municipalities, staff in the land use planning, building permits and inspections areas can provide valuable information on regulations that guide a wide range of landscape elements. Examples include the maximum height of fences, location of driveways or places where visual obstructions to moving vehicles, like hedges, should be avoided. Even if they seem constraining, regulations can prevent disputes between neighbours about fences, undesirable views, and a host of other issues. In certain neighbourhoods, especially heritage districts, bylaws may be very strict and may go as far as prohibiting certain materials and defining the landscape style.

Some municipalities regulate tree removal, even on private property, and require a permit. The goal is generally to ensure that neighbourhoods enjoy the benefits of having trees. Some also enforce watering restrictions and regulate pesticide use.

Municipal right-of-ways

Municipalities own and manage a large portion of the land in the neighbourhood, like parks and the road network. The municipal right-of-way beside the street includes not only sidewalks, but also water, sewer, hydro and other utilities. To allow room for this infrastructure, the right-of-way is usually larger than the pavement and sidewalk. In many municipalities, homeowners are responsible for landscaping this space (subject to certain restrictions) and maintaining the areas within the right of way adjacent to their properties, even though it is not legally theirs. Keep in mind that this area may be dug up for infrastructure repairs, and the municipality may not be responsible for reinstating damage to the landscape features you have added.

Property limits

The property limit is the place where your property ends. If the front boundary borders on the municipal right-of-way, there are still three neighbours in your back and side yards. Landscape work near or at the property limit could have an impact on them. Your design could be influenced by their concerns for privacy, safety, drainage, aesthetics and enjoyment of their own property. Regulations regarding each person's obligations with respect to this boundary vary from one community to another, so verify with your municipality.

Covenants and easements

Covenants are conditions that are registered with the legal description of the property. Although owners may change, the conditions in the covenants remain with the property. Condominium associations, as well as the adjacent owners of semi-detached houses may have covenants over your property. An easement is a right of passage granted to another through your land. More frequently, the electrical, gas or telephone suppliers use underground or aerial easements to allow the passage of overhead cables or buried conduits that supply the neighbourhood. These easements give them the right to install their equipment on a homeowner's land and to access it for maintenance or repair. Conservation agencies may also hold easements or covenants on portions of your land, such as trail routes or stream banks. Owners' rights are limited in these areas. They are usually indicated on your lot survey.

Building codes

A building permit may be required for certain landscape structures, such as decks and stairs, to meet building code requirements. Contact your municipal building department for information on the requirements.

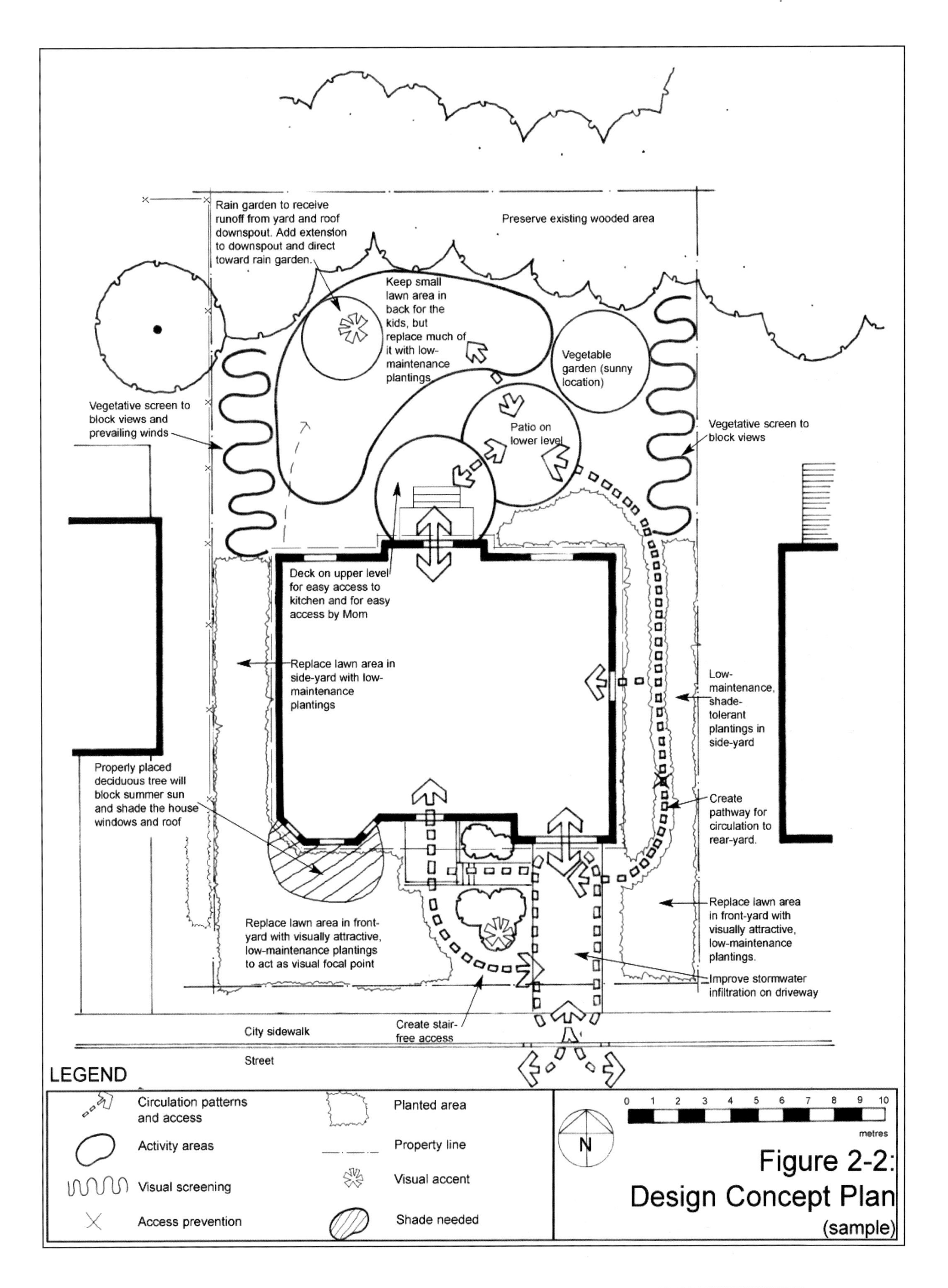

Figure 2-2: Design Concept Plan (sample)

Environmental regulations
Approvals may be required by federal, provincial, territorial and municipal governments when undertaking work that may have an impact on the environment. For example, various agencies have regulations and require approvals for work to be undertaken by a body of water, including your local municipality, the federal Department of Fisheries and Oceans, and provincial and territorial ministry responsible for the environment or natural resources. Also contact your municipality, province or territory for regulations that may apply to projects that affect natural features such as environmentally sensitive areas, wetlands, significant woodlands and other important ecological features.

Again, some municipalities have restrictions on the use of pesticides and tree protection bylaws. Provincial legislation regulates certain elements like the control of noxious weeds.

Check before you dig.
You are responsible for repairs if you damage underground utilities when you are digging. So when in doubt, phone the relevant agency or company to locate the utilities, like hydro, cable, phone, gas, water and sewer.

Step 5. Design the Layout (Preliminary Design Plan)

The purpose of the preliminary design plan is to refine your bubble diagram or concept plan. Now that the preferred locations for activities and landscape features have been chosen, you can define their shape and materials (Figure 2-3). Just like with the site inventory and analysis plan, you should be working on a plan that is to scale. Once the plan has been completed there will still be many decisions to make. The following chapters may help you make those decisions.

Step 6. Work Out the Technical Approach and Choose the Materials

The chapters that follow provide many tips on choosing the materials, including initial costs, durability, maintenance considerations and environmental pros and cons. They also illustrate and discuss technical issues, such as installation methods. Many publications, Web sites and other information sources can supply technical tips as well. Garden centres, nurseries and suppliers or manufacturers of building materials can provide you with product information, costs and installation methods.

Choosing healthy materials

- Favour materials that are harvested or manufactured locally to help reduce transportation energy.
- Favour products containing recycled materials or those made from renewable or recyclable resources.
- Use products destined to be reused or recycled at the end of their service life.
- Use products that are not associated with endangered species (habitat destruction).
- Think about the degradation of soils, water, air and habitats caused by mining operations, or the removal or extraction of raw materials.
- Avoid materials that produce a lot of hazardous waste or that are manufactured using hazardous substances.
- Avoid products and finishes that contain toxic or hazardous substances.
- Select durable products that require little maintenance.
- Reduce, reuse or recycle construction wastes, whether they are materials or packaging.
- Think about the energy inherent in construction materials; use products that require less energy in the manufacturing process.
- Favour materials that allow water to soak into the soil and reach the water table.

Step 7. Estimate Costs and Compare Them to Your Budget

At this stage, assess the costs of supplying and installing the various elements of your design. These costs, compared to the initial budget, will guide your selections. For example, they will enable you to choose between two materials for the same surface. Elements that are too costly can perhaps be eliminated or replaced. Phasing, or spreading the project out over a number of years, may also become a possibility or even a necessity. While this book provides initial costs of different materials relative to the others, it does not quote exact amounts. For more specific costs, consult

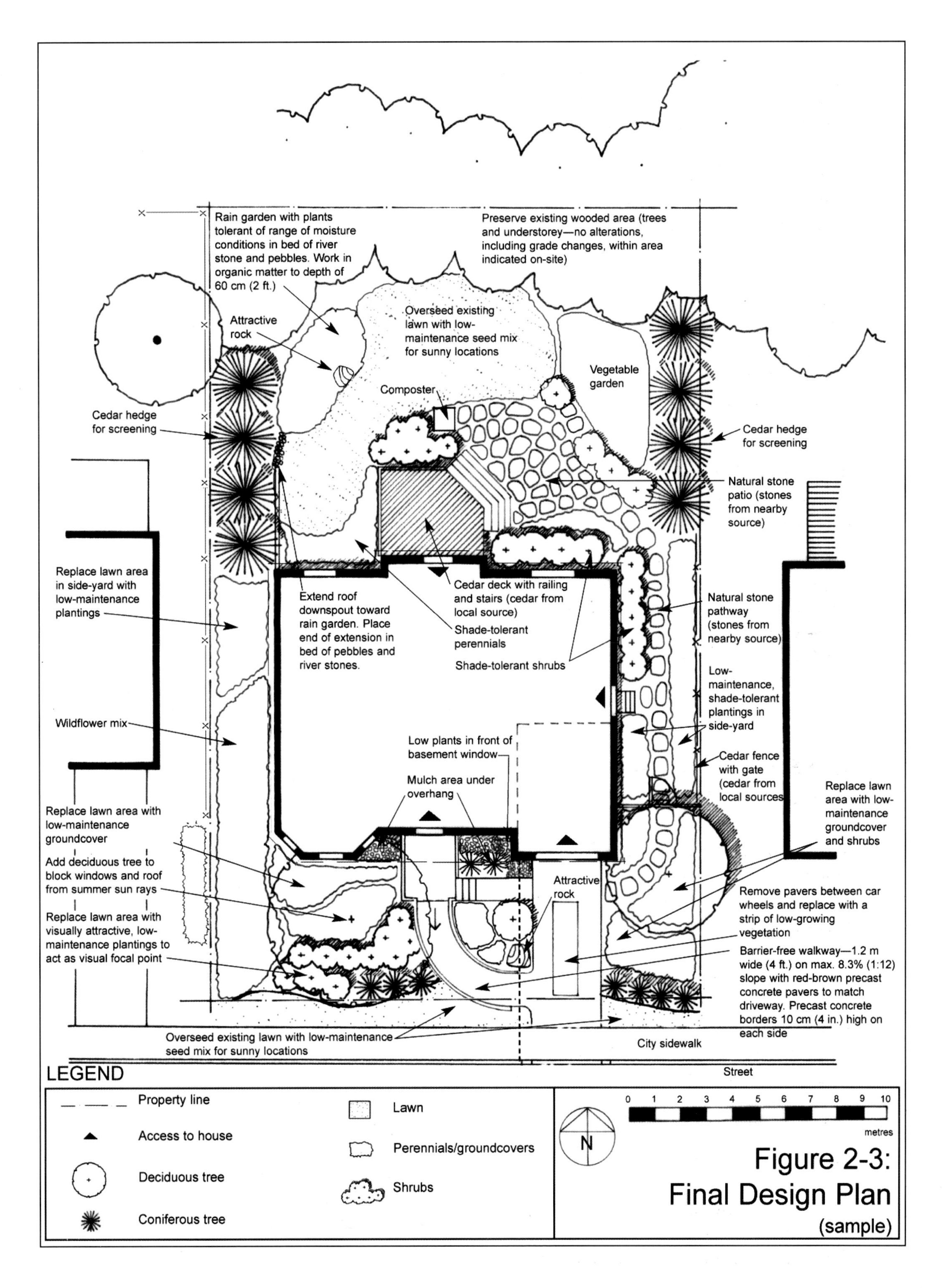

Figure 2-3: Final Design Plan (sample)

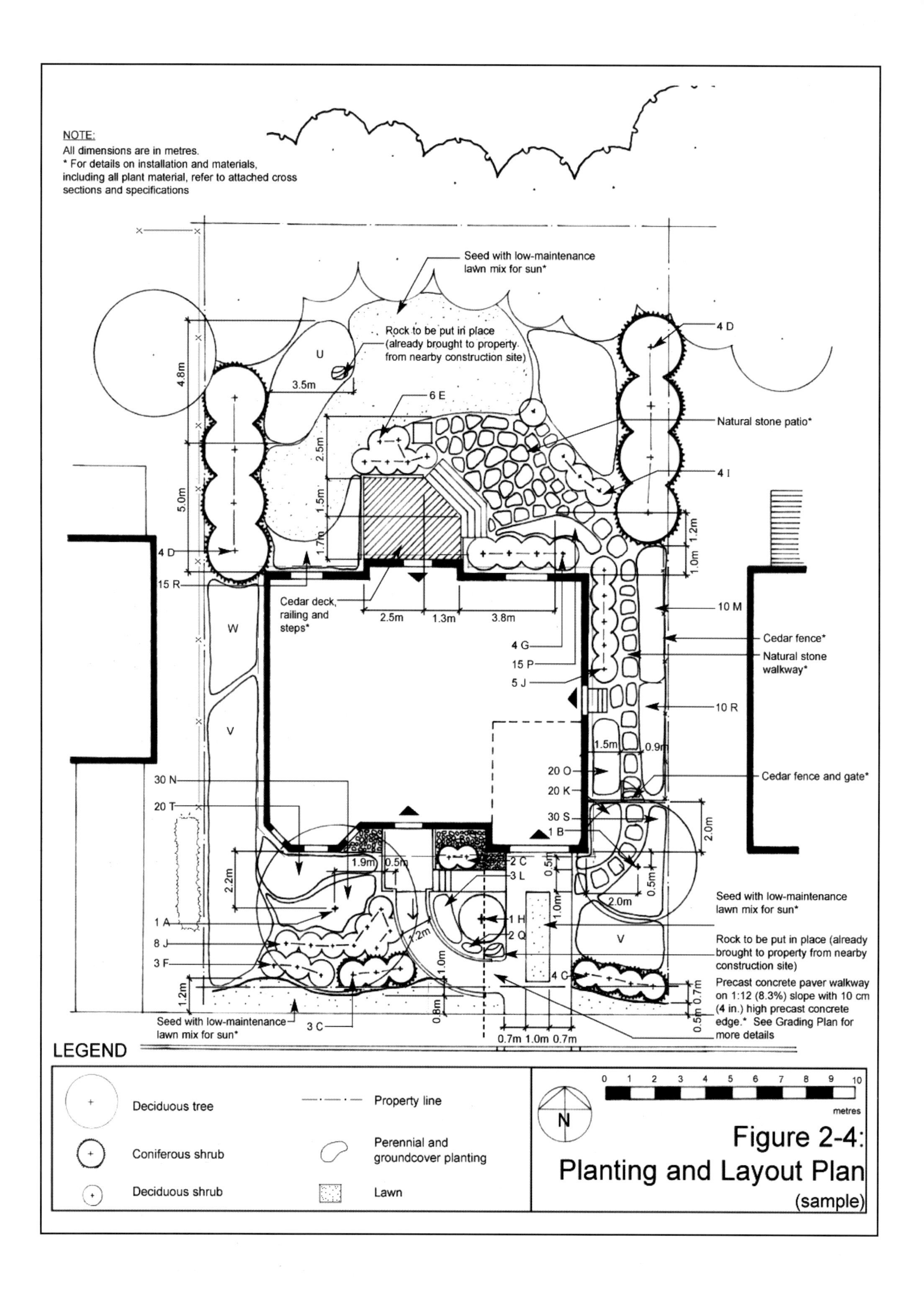

Figure 2-4: Planting and Layout Plan (sample)

Plants for Figure 2-4: Planting and Layout Plan

Code	*Latin name* / Common name
	Deciduous Trees
A	*Fraxinus pensylvanica* / Green, Red Ash
B	*Sorbus americana* / American Mountain Ash
	Shrubs
C	*Juniperus horizontalis* / Creeping Juniper
D	*Juniperus virginiana* / Eastern Redcedar
E	*Aronia melanocarpa* / Black Chokeberry
F	*Diervilla Lonicera* / Low-bush Honeysuckle
G	*Dirca palustris* / Leatherwood
H	*Physocarpus opulifolius* / Ninebark
I	*Potentilla fruticosa* / Shrubby Cinquefoil
J	*Symphoricarpos albus* / Snowberry
	Perennials
K	*Actaea rubra* / Red Baneberry
L	*Elymus Canadensis* / Canada Wild Rye
M	*Matteucia struthiopteris* / Ostrich Fern
N	*Hepatica acutiloba* / Sharp-lobed Liverleaf
O	*Polygonatum multiflorum* / Solomon's Seal
P	*Polystichum acrostichoides* / Christmas Fern
Q	*Sedum spectabile* / Showy Stonecrop
R	*Smilacina racemosa* / False Solomon's Seal
S	*Tiarella cordifolia* / Foamflower
T	*Geranium maculatum* / Wild Geranium
U	Rain garden mix (tolerate range of moisture conditions): *Andropogon scoparius* / Little Bluestem (3) *Aronia melanocarpa* / Black Chokeberry (2) *Aster novae-angliae* / New England Aster (5) *Hypericum prolificum* / Shrubby St. John's-wort (3) *Heliopsis helianthoides* / Sweet False Sunflower (3) *Solidago canadensis* / Canada Goldenrod (4) *Viola canadensis* / Canada Violet (10)
V	Native wildflower meadow mix (sun) - 9 of each for west garden, 6 of each* for east garden: **Asclepias tuberosa* / Butterfly Weed **Andropogon scoparius* / Little Bluestem **Echinacea purpurea* / Purple Coneflower *Elymus Canadensis* / Canada Wild Rye **Monarda didyma* / Bea Balm *Monardia fistulosa* / Wild Bergamot *Oenothera biennis* / Evening Primrose **Panicum virgatum* / Switchgrass *Rudbeckia hirta* / Black-eyed Susan *Solidago canadensis* / Canada Goldenrod
W	Shade-tolerant mix - 8 of each: *Actaea rubra* / Red baneberry *Polystichum acrostichoides* / Christmas Fern *Smilacina racemosa* / False Solomon's Seal *Solidago flexicaulis* / Zig Zag Goldenrod

Plant list: also indicate condition, size and total quantities of each species, as in the sample plant list in Chapter 6. Check plant catalogues or labels for information, like correct spacing.

product catalogues that may be available at local nurseries or building material retailers or call suppliers directly. Chapter 8 provides a process for obtaining cost estimates from contractors.

Even if they are more expensive initially, more durable materials can often save you money in the long run.

Long-term costs should also be factored into your decisions. The book also discusses and compares the durability and maintenance considerations of different materials and approaches. Even if they are more expensive initially, more durable materials can often save you money in the long run.

Step 8. Modify the Design to Create Your Final Design Plan and Technical Drawings

Figure 2-3 shows an example of final landscape plan. Taking technical characteristics and costs into account will inevitably lead to certain modifications of the preliminary plan. It is entirely normal for the project to evolve as you discover the benefits and constraints of different materials. The adjustments you make will improve the quality of the project and further adapt your concepts to the site. Input from professionals may result in further refinements to your final landscape plan. You may also need to make further adjustments after seeking the necessary permits and approvals.

Even if you have prepared the preliminary design plan, you may want to seek professional help on technical matters, for example from a landscape architect, including preparation of detailed specifications and technical drawings. These provide precise direction on installation methods and other technical issues. If you are hiring a contractor to do the landscape work, you'll need to clearly identify the layout, plant materials, grade changes and materials you are proposing in your design, for example through planting plans, layout plans, grading plans, technical cross-sections and specifications. Figure 2-4 shows a planting/layout plan and Figure 3-3 shows a grading

Low-cost ideas for children:

- Define an area for a garden that kids can create themselves, and grow plants that they find interesting, like pumpkins. For example, a ring of sunflowers could mark out a circular play space. Access to their special area could be defined by a temporary gateway that is sized just for them, like one made of pruned or fallen branches that are bent and interwoven.
- Give them annual seeds, like carrots or lettuce, they can sow into interesting shapes.
- Avoid plants with toxic parts or those with thorns.
- Place narrow, windy foot paths in your planted areas, like tall grasses and wildflower gardens or woodland shade gardens.
- Place plants that attract birds and butterflies.

SOILS AND GRADING

Soil is an important element of landscape design, yet it is often the most ignored. Out of sight, out of mind. Furthermore, many residential developments are characterized by compacted, poor soil. So soil improvements are often necessary. Without proper soil structure, pH, fertility, moisture and texture your plants may not thrive.

However, preserve existing soil to the extent possible by keeping it in place, tailoring your plant selections to your soil conditions and making only minor amendments where necessary, rather than removing and replacing it with new material, like topsoil carted in from other sites. This will save you money, reduce energy required in transporting the materials as well as preserve the plant and life forms that already exist in the soil. But the extent of your soil improvements will depend on the quality and health of soil already existing on your site and on the soil preferences of the plants you select. This chapter will help you analyze your soil so that you can select the plants that suit it, and help you make only necessary amendments needed for the health of your plants.

Preserving what's already there also applies to grading, the process of reshaping the topography or ground level to accommodate your home, and features around it, for example, trees, patios, pathways, driveways and more. It also ensures that water goes where you want it. Plan your grading in parallel with the landscape design. While a certain amount of grading is usually needed to accommodate your features and ensure proper drainage, plan your grading to minimize changes to what you already have. This chapter provides grading tips and is followed by a related chapter, which deals with stormwater and water conservation.

SOIL

Healthy soil is the foundation of healthy plants. It not only holds plants in place but also provides the nutrients, water and air needed to support them. It contains a host of micro-organisms, insects and fungi that participate in the plant growth process. Relatively undisturbed soil occurs in layers or "horizons" (figure 3-1). Some micro-organisms, insects and fungi live solely in the top centimetres of soil and others live deeper. Where you have natural or undisturbed soil on your property, avoid over-mixing the layers, which can bury organic matter and other healthy soil organisms too deeply and affect other soil conditions like moisture and texture. In developed urban areas, the soil profile usually looks quite different because of backfilling and mixing of layers.

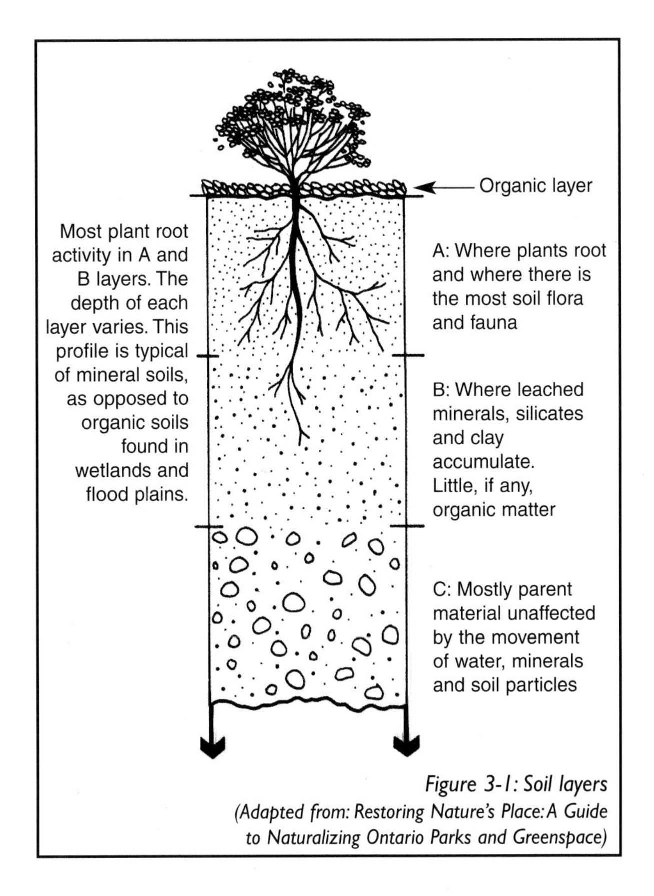

Figure 3-1: Soil layers
(Adapted from: Restoring Nature's Place: A Guide to Naturalizing Ontario Parks and Greenspace)

Get to Know Your Soil

Each plant has its own soil needs. You can find out about these specific needs by asking at your local nursery and looking in specialty books and plant catalogues. They will describe each plant's needs in terms of drainage, moisture, fertility, pH and texture. For example, they may tell you that white pines prefer sandy loam soil. If you have clay soil, you could try to amend it, but the cheaper, easier and more environmental solution is to choose a plant that is more adapted to that soil condition. That is why it's best to know your soil type from the outset.

There are regional soil maps that broadly show the soil types of a given region. It is difficult to rely solely on these maps as several factors may influence the soil composition on your individual site. Even within your own property, the soil characteristics may vary according to whether you are in a swale bed, on a slope or on a peak or whether the soil has been significantly altered, for example by construction.

To know your soil, use the eight following tests:

1. Humus and organic matter

Humus is produced by the decomposition of organic matter, such as leaves. It is an essential feature that determines the soil's nutrient content or fertility but also affects whether it's easy to work with or compacted, its ability to hold or drain water and more. In a forest, humus is created naturally by decomposition of leaves, bark and other organic matter. Everything relates to a cycle. A tree loses its leaves in the fall. The leaves then decompose on the ground. The humus created releases nutrients, which are then absorbed by the tree to make new leaves. Different plant species have varying requirements for humus. It's best to select plants that are suited to your soil's fertility, for example, if you have infertile soil with a low humus content, select plants that tolerate those conditions.

Humus is only available through organic matter, which is available in many forms, like compost, manure, grass clippings, leaves and more.
For many of you, this is a resource that you can get free of charge, right on your own property.

Test: Is there a layer of humus and how thick is it?
Dig a small pit to a depth of at least 30 cm (1 ft.). Hardened, compacted layers of soil are an indicator of little biological activity, inadequate drainage, low fertility and low humus content. The blacker the soil, the more humus it contains and the more fertile it is.

2. Soil texture

Soil textures or types are mainly classified according to the relative proportion of sand, silt or clay. Classification is made based on the diameter of the particles; sand particles are large and clay particles are fine. Texture influences nutrient content, moisture and drainage in several ways. Clay particles, like humus are negatively charged. They attract positively charged water molecules and metal ions. So clay soils are usually rich in minerals, but are hard to drain because they retain water particles. Sandy soil drains easily, and can result in a dry environment that lacks fertility as the sand particles do not adhere to the water and metallic ions.

Loam has a mixture of sand, silt and clay particles and also contains humus. Loamy soil is ideal for most garden plants because it holds plenty of moisture but also drains well and is fertile and easy to work or friable. There are, however, many plants that are tolerant of a variety of soil types, while others have more specific soil type requirements. The best approach is to choose plants that are suited to your soil type. There are a number of intermediate classifications of soil texture such as loamy sand or sandy loam.

Several simple tests make it possible to identify your soil texture as shown in Table 3-1.

Table 3-1: Testing Soil Texture

Texture	Feel Test	Moist Cast Test	Ribbon Test	Shine Test
sand	grainy, little floury material	no cast	can't form a ribbon	
loamy sand	grainy with slight amount of floury material	very weak cast, does not allow handling	can't form a ribbon	
silty sand	some floury material	does not allow handling	can't form a ribbon	
sandy loam	grainy with a moderate amount of floury material	weak cast, allows careful handling	barely forms a ribbon, 1.5 - 2.5 cm (0.6 - 1 in.)	
loam	fairly soft and smooth with obvious graininess	good cast, easily handled	thick and very short, <2.5 cm (1 in.)	
silt loam	floury, slight graininess	weak cast, allows careful handling	makes flakes rather than a ribbon	
silt	very floury	weak cast, allows careful handling	makes flakes rather than a ribbon	
sandy clay loam	very substantial graininess	moderate cast	short and thick, 2.5 - 5 cm (1 - 2 in.)	slightly shiny
clay loam	moderate graininess	strong cast clearly evident	fairly thin, breaks easily, barely supports its own weight	slightly shiny
silty clay loam	smooth, floury	strong cast	fairly thin, breaks easily, barely supports its own weight	slightly shiny
sandy clay	substantial graininess	strong cast	thin and fairly long, 5 - 7.5 cm (2 - 3 in.) holds its own weight	moderately shiny
silty clay	smooth	very strong cast	thin and fairly long, 5 - 7.5 cm (2 - 3 in.) holds its own weight	moderately shiny
clay	smooth	very strong cast	very thin and very long, >7.5 cm (3 in.)	very shiny

Source: Field Manual for Describing Soils in Ontario

Feel tests

Graininess: Thoroughly dry and crush a small amount of the soil by rubbing it with the forefinger in the palm of your other hand. Then rub some of it between your thumb and fingers to measure the percentage of sand. The more grainy it feels, the higher the percentage of sand.

Stickiness*: Compress wet soil between the thumb and forefinger. Determine the degree of stickiness by noting how strongly it adheres to your thumb and forefinger, and how much it stretches when you release the pressure.

Moist cast test* (Figure 3-2)

Compress moist soil by squeezing it in your hand. When you open your hand, if the soil holds together (for example, forms a cast), pass it from hand to hand; the more durable the cast is, the higher the percentage of clay.

Ribbon test* (Figure 3-2)

Roll a handful of moist soil into a cigarette shape and squeeze it between your thumb and forefinger to form the longest and thinnest ribbon possible. Soil with high silt content will form flakes or peel-like thumb imprints rather than a ribbon. The longer and thinner the ribbon, the higher the percentage of clay.

Shine test

Form a ball of moderately dry soil and rub it once or twice on a knife blade. The more shine, the higher the percentage of clay.

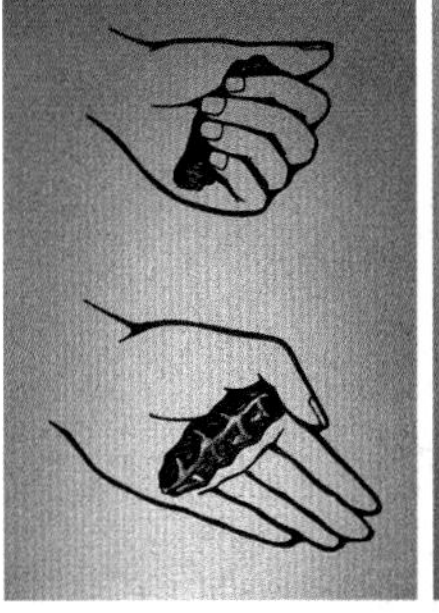

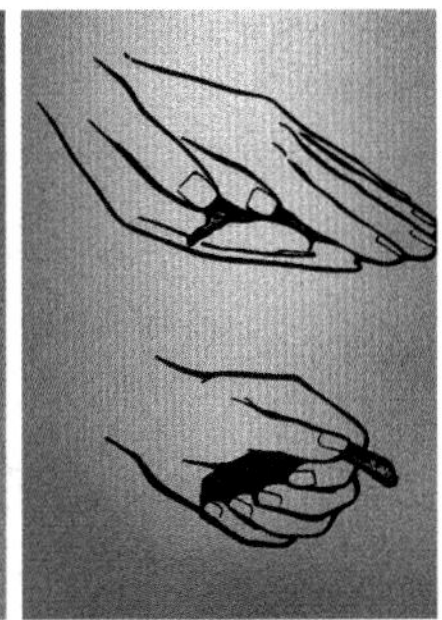

Figure 3-2: Testing soil texture
The moist cast test (left), the ribbon test (right)
Source: Regional Municipality of Durham

*For each of these tests the soil specimen should be gradually moistened and thoroughly reshaped and kneaded to bring the specimen to its maximum "plasticity" and to remove dry lumps. Do not add too much water to the point where the specimen becomes "sloppy" and begins to lose its coherence.

3. Soil structure and porosity

Structure refers to the size and arrangement of the particles. It determines porosity or the presence of air between the particles. The small spaces between soil particles are where the plant roots pick up the oxygen, nutrients and water they need to grow. Soil with good structure is porous and may contain up to 25 per cent air. It also contains a balance between large particles that allow drainage and fine particles that retain nutrients and moisture. Structure and porosity are also affected by the presence of earthworms, insects, bacteria, fungi and other micro-organisms. Porosity is also a factor that allows the exchange of gases, like oxygen, among roots, micro-organisms and the atmosphere. Compaction affects porosity, making the soil difficult to work and hard for roots to grow freely and for plants to take up water, oxygen and nutrients.

Test: Is your soil compacted?

You can spot compacted soils visually. Bare, plantless patches of soil have probably been compacted. Certain plants are typical of compacted soils, for example: English daisy, field bindweed, plantains, quack grass and dandelion. Areas that puddle could also indicate compaction.

Digging the soil will also help determine how compacted it is. If your shovel penetrates it easily, that means it is porous and its structure is good. When you remove a shovelful of soil, it should crumble easily.

4. Moisture

Moisture levels are a function of local climate, topography and the soil characteristics already discussed. They can therefore vary greatly from one place to another in a single region and even on a single site. Moisture has a direct influence on the plant species your soil can support.

The degree of soil moisture is often described as wet, moist or dry. Some plants are adapted to a very specific level of moisture within the range, while others can adapt to significant fluctuations, and still others can tolerate moderate fluctuations. It's best to choose plants that are suited to the soil moisture.

Test: Check moisture levels

Look at your property at different times of the year to locate the areas where water accumulates and where it drains rapidly. Identify these areas on your site inventory plan. Depressions or low areas are generally wetter than high areas or slopes. Fine soils like clay hold water more than course ones, like sandy soils. For specific levels at any given time, use the soil texture-feel tests noted above. Sunny locations tend to be drier than shady locations.

Another indicator of soil moisture is the plant species that thrive in different locations, according to their specific moisture needs. Also, ground vegetation that is thick and lush may indicate moister soils than where it is sparse and thin.

Organic matter improves soil fertility, structure and moisture. It also increases micro-organism population. In sandy soil, it improves water and nutrient holding capacity, and in clay soil it helps improve aeration and drainage. All this results in healthier plants. Only organic matter can supply humus. Good low-cost sources of organic matter are compost, manure, grass clippings and leaves.

5. Fertility

Fertility refers to the soil's ability to store and release the nutrients required to allow healthy plant growth. The main nutrients are nitrogen, phosphorus and potassium (N, P, K) and there are trace elements like calcium and magnesium.

Fertility is linked to soil texture, humus, organic matter, and the micro-organisms present in the soil, which break down and recycle organic matter into nutrients that are capable of being absorbed by the plants.

Some species like those on the forest floor require rich soils while prairies or meadows are happy with less fertile soil. Soils with a thick layer of humus and a good thick A horizon are fertile. Intensive cultivating of the soils accelerates the decomposition process and reduces the quantity of organic matter, making them less fertile.

Soils containing clay are usually more fertile than sandy soils because the small clay particles adhere more rapidly to certain nutrients than the larger grains of sand or silt.

Test: How to measure fertility
Chemical tests are the most effective test for soil fertility. They detect the composition of the soil and the elements it contains, such as N, P, K. See below for more information on chemical tests.

6. pH

The pH measures the soil's acidity or alkalinity. Most plants do well in soil with a neutral pH of 6.6 to 7.4, or in moderately acid soil with a pH of 6.0 to 6.5. However some plants require very specific pH levels. Alkaline conditions promote the rapid decomposition of organic matter, resulting in a rich, thick layer of soil typical of forests. Soils with a granite base typical of the Canadian Shield tend instead to be acidic.

In urban environments, surfaces like concrete and gravel, in addition to salt spray and other factors, can change the soil's natural pH. So do not rely on regional data for information about soil pH on your specific property.

Test: How to evaluate soil pH

- Check your property for plants that grow in an acidic, neutral or alkaline soil.
- The best way to test soil pH is to perform a chemical test.

7. Soil depth

Soil depth refers to the depth of the soil above bedrock. Deep soils—over 75 cm (30 in.) deep— can support greater varieties of plants. However, several plants have adapted to shallow soils. If unsure, check to confirm suitable depths before selecting species.

8. Life in the soil

Healthy soil contains specialized micro-organisms that are very active. They are responsible for, among other things, the decomposition of organic matter, such as leaves, to turn them into humus. During this process, minerals and essential elements are released and absorbed by the plants. Many micro-organisms and fungi have developed relationships with plant roots by helping them absorb the nutrients. Their absence from the soil sometimes explains why certain plants cannot be transplanted or cannot establish themselves in a given environment even if all the other survival conditions are met.

The larger creatures, like worms and millipedes, aerate the soil as they burrow tunnels and mix the soil, preventing it from becoming too compacted. A properly aerated soil allows for better exchange of nutrients. Improper use of pesticides can eliminate beneficial life forms, including those that aerate the soil.

Test: Is there life in your soil?
Look for the various insects and worms in your soil. The more of them, the healthier your soil. To see them better, dig holes and study the soil you lifted up.

Chemical Tests
To determine the soil's characteristics, like the fertility (N, P, K and other nutrients), texture (proportion of sand, silt and clay) as well as pH and other factors, a chemical analysis may prove necessary to provide accurate information. A number of laboratories do this type of testing. Contact a local garden centre, soil testing laboratory or provincial agriculture ministry for more information like how to take a soil sample and where to bring it. Medium- and low-cost home testing kits are also available.

How to Amend Your Soil

After doing these tests, you now know what soil you have in various locations on your property. Now you can choose plants that are best suited to your soil. Ask at your local nursery or look in the many plant catalogues available at your local nursery and the several specialty books at the library or bookstore to help you determine the soil needs of the plants you are considering using. The plant list in this publication may also be helpful.

However, there are situations in which amendments may be needed. These include: areas in which soil health has been seriously compromised; food-producing gardens or other specialized plantings that may not be as easily tailored to your soils; and, areas in which even well-suited plants are not thriving. Amendments can also be helpful when establishing new plantings. Having plants that are well suited to your soil can save you time and money down the road by avoiding plant replacement and extra maintenance, but if you do need to amend the soil, do so selectively in specific locations only as needed.

Table 3-2 shows frequently encountered soil deficiencies and solutions. Whatever amendments you make, they should be worked into the soil, rather than leaving them on the surface. This will help spread the improvements around, prevent drying out, particularly of organic matter and avoid harmful concentrations. This will also help to encourage the roots to spread, rather than concentrate only in the area of amended soil.

The area you want to plant in may consist of such poor quality material that you find you need to replace it with a more appropriate growing medium, such as loam soil and organic material, like compost. An example is the gravel or crushed rock base of an area where you are replacing pavement with plantings. But keep in mind that there are plants that will grow in gravel.

Compost
To improve soil fertility, structure, moisture, humus and other factors, compost is an excellent soil amendment. It's a recycled, renewable, locally available resource you can produce at home. Look for the composting tips in Chapter 7. Many municipalities offer low-cost compost from yard waste composting.

The addition of 5.0 - 7.5 cm (2 - 3 in.) of compost mixed into your soil to a depth of approximately 20 cm (8 in.) is a common soil amendment that addresses several soil problems.

Table 3-2: Soil problems and solutions

Soil issue	Problems	Solutions
high clay content	• compacted, heavy and difficult to handle • insufficient drainage, crust formation in dry weather • difficult for plants to root, establish and grow	• select plants suited to clay soil • work in organic matter, like well rotted manure and compost • aerate • plant green manure* • work in coarse sand
high sand content	• soil generally low in organic material and less fertile • good drainage, poor water-retention capability	• select plants suited to sandy soil • work in organic matter, like well rotted manure and compost
poor structure	• difficult to work the soil • poor drainage • inability to hold nutrients • lack of fertility and aeration • plants experience growth difficulty	• select plants suited to poor soil structure • work the soil and add in compost, manure or leaves • aerate • plant green manure*
poor soil moisture	• plants experience growth difficulty and wither	• select plants suited to dry soil • work in organic matter, like compost, manure or leaves
low fertility	• low levels of organic material and micro-organisms • poor drainage and structure, hard to work, • inability to hold nutrients • plants experience growth difficulty	• select plants suited to low fertility • in spring or fall, work in well-rotted organic matter like manure, compost. You can also incorporate grass clipping and leaves.
low nitrogen	• plant growth problems • pale green or yellow foliage	• work in compost, manure, grass clippings, leaves. Other natural sources of nitrogen include blood meal, corn glutenmeal and fish emulsion. • plant nitrogen fixing species • plant green manure*
low potassium	• plant growth problems • poor stem strength • scorched leaf edges • less disease resistance and winter hardiness	• work in sources of potassium such as ground rock potash (granite dust), wood ash, compost, dried poultry manure, leaves or hay
low phosphorus	• plant growth problems • late maturity • purple colour on the leaves	• work in sources of phosphorus such as bone meal or rock phosphate
too alkaline	• soil nutrients less available • plants experience growth difficulty	• select plants suited to alkaline soil • work in powdered sulphur, coniferous needles, composted woodchips and sawdust to lower pH
too acidic	• soil nutrients less available • plants experience growth difficulty	• select plants suited to acid soil • lime, wood ash, and bone meal raise pH • make adjustments gradually based on annual testing results

* Green manure consists of herbaceous plants that improve site conditions for subsequent plants, for example, soil-enriching plants that fix nitrogen. They are grown on the site temporarily, cut and worked into the top 15 cm (6 in.) of soil before they flower and go to seed.

GRADING

As discussed earlier, grading is the process of reshaping the topography or ground level to accommodate your home, and features around it like trees, patios, pathways, driveways and more. It also ensures that water goes where you want it to go.

Why grade?

- to drain surface water, or runoff, away from the house
- to manage surface water effectively on your site (more details in Chapter 4)
- to integrate your activities, like play areas, into the site
- for aesthetic reasons, like defining spaces, screening or creating views or controlling wind and noise
- to integrate the house into the site
- to help conserve existing features, such as trees

For more details on stormwater, refer to Chapter 4.

Grading changes are generally planned and described on a technical plan called the grading plan, used to calculate the vertical changes you want to make. (Figure 3-3)

Grade changes can also be added to your landscape plan. You will have already collected most of the relevant topographical data when you did your site inventory and analysis (see Chapter 2).

One way to show topography is to use spot elevations. These are indicated at certain critical places on the grading plan, such as at the base and top of retaining walls and stairs. It's the simplest method to identify site levels.

Useful site information for grading

- high and low points
- levels at corners of your lot along property boundaries including along the street, roads, sidewalks and curbs
- outdoor levels at corners of your house
- indoor levels at your house doorways
- top and bottom of special features, such as berms, patios, steps and decks
- main shapes and high point of special topographical features, such as rock outcroppings, ponds
- drainage features, such as storm sewers, ditches and swales
- direction of surface drainage and slope gradient at important locations; for example, to show how water is draining away from the house
- levels under existing trees
- soil characteristics that affect drainage, like soil texture and structure; for example, compacted or clay soil has a lower infiltration rate than uncompacted or sandy soils
- when designing an area for stormwater infiltration, refer to the section on "How to determine the size of a rain garden" in Chapter 4 which lists the site features you should analyze. For example, you'll need to estimate the volume of stormwater runoff from your roof as well as the percolation rate of your soil.

Another method for showing ground levels on the grading plan is contour lines. These are lines representing a constant level on the surface of the ground. The lines are used to indicate the changes between the existing topography and the proposed grading. Depending on the plan's scale, the contour lines are usually indicated at vertical intervals of 0.5 m (1½ ft.). The existing lines are normally indicated by broken lines, and the proposed contour lines are indicated by a solid line. A combination of elevations and

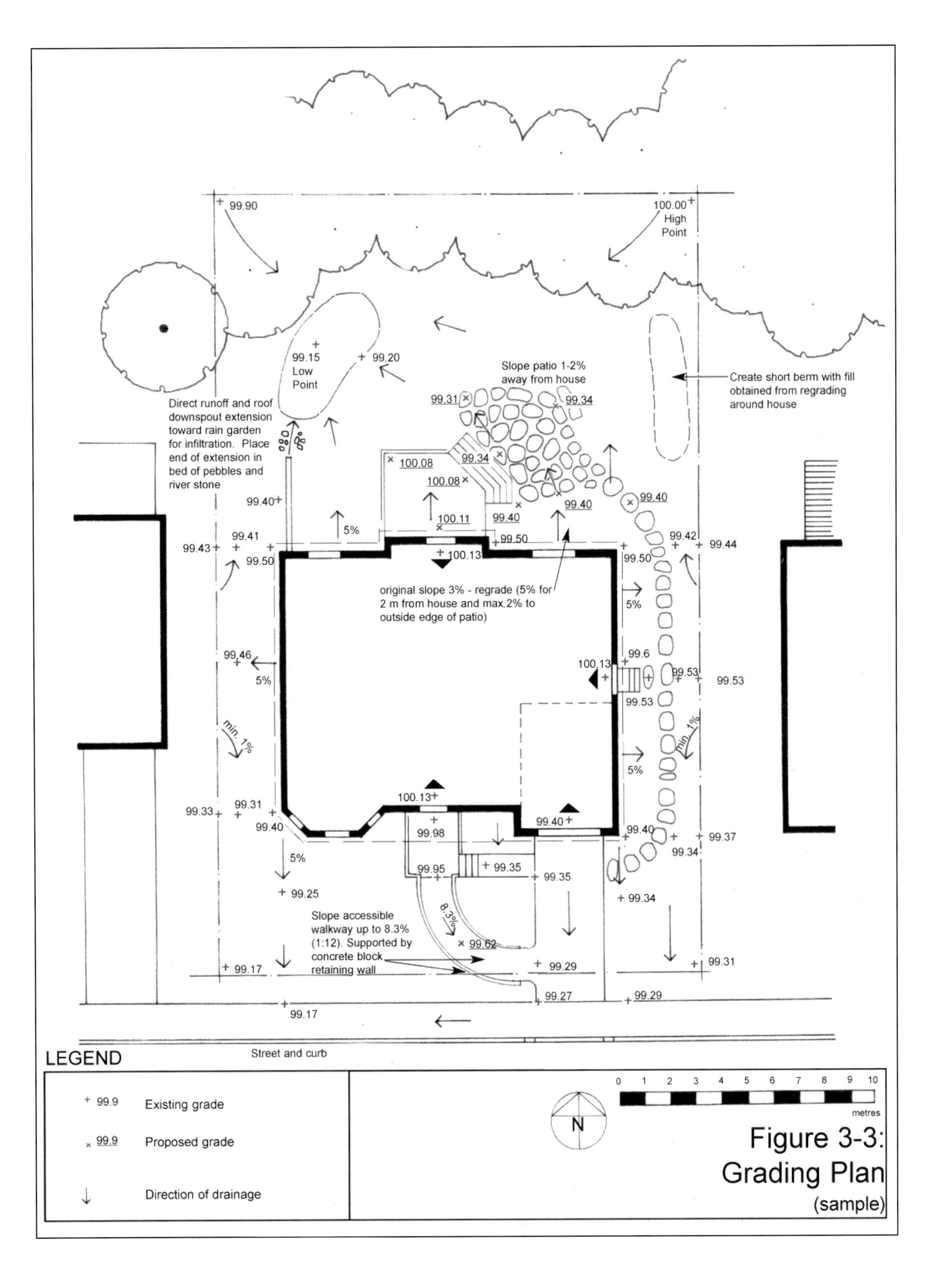

Figure 3-3: Grading Plan (sample)

contour lines can be used to clearly describe the slopes and shapes of the ground surface.

The slope is the ratio of the rise (the vertical change) to the length of the run (the horizontal change); for example, 1 in 10 represents a 10 per cent slope—a 1 m (3 ft. 3 in.) rise over 10 m (32 ft. 10 in.). One in 50 represents a 2 per cent slope—a 1 m (3 ft. 3 in.) rise over 50 m (164 ft.). The same 2 per cent slope could also be expressed as 1:50 or 50:1.

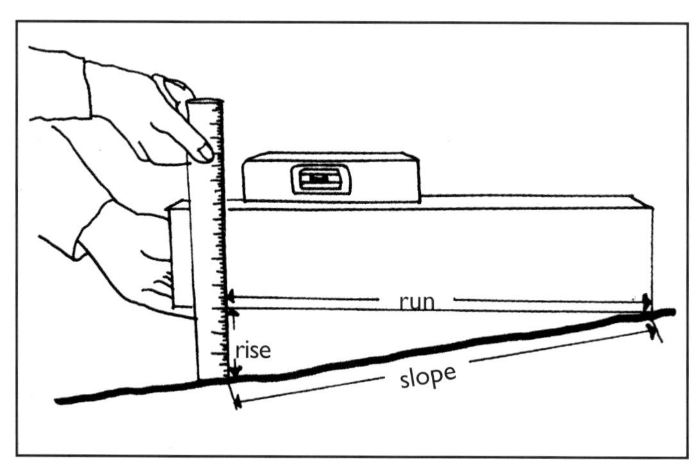

Figure 3-4: How to measure a slope
The slope is the ratio of the length of the rise (the vertical change) to the length of the run (the horizontal change). A simple way to measure slope is to use a level placed on a 2x4. Place the board on the ground along the slope you want to measure and lift the lower end up until the board is level. Measure the distance from the ground to the bottom edge of the board at the end of the slope you want to determine. That is the rise and the length from the end of the board to where you measured is the run. Divide the rise by the run to obtain the slope. For example, if the rise is 5 cm (2 in.) and the run is 2.5 m (8 ft. 2½ in.), the slope is 2 per cent.

Check and adhere to your municipality's standards. Slope the ground away from your house around the entire perimeter. While at least a 5 per cent slope away from foundation walls is acceptable, a 10 per cent slope for the first 2 metres is preferred to ensure the correct slope after the soil settles and compacts. Use at least a 2 per cent slope for impermeable surfaces adjacent to the house, such as a driveway or patio. For more tips on stormwater and drainage, including proper drainage around the house and infiltration, refer to Chapter 4.

Figure 3-5 shows characteristics of different gradients. Generally, patios and decks should slope 1 to 2 per cent. The suitable gradient for walkways and driveways is particularly important during winter's slippery conditions. In general, for relatively flat areas that experience snow and ice, it is suggested that the crossfall (along the width of the driveway or path) not exceed 2 per cent and the longitudinal slope (along the length) not exceed 7 per cent. Again, this will depend on local conditions, so check local municipal standards. For barrier-free access, the maximum slope should be 5 per cent, unless it is designed as a ramp, which generally can have a slope of up to 8.3 per cent (1:12). When a slope you need to walk on is so steep that walking is difficult, consider installing steps or ramps. For more information on walkways, patios and driveways, refer to Chapter 5.

1. Preserve Your Property's Topsoil

Before doing any excavation or grading work, strip off the topsoil and reuse it for plantings on

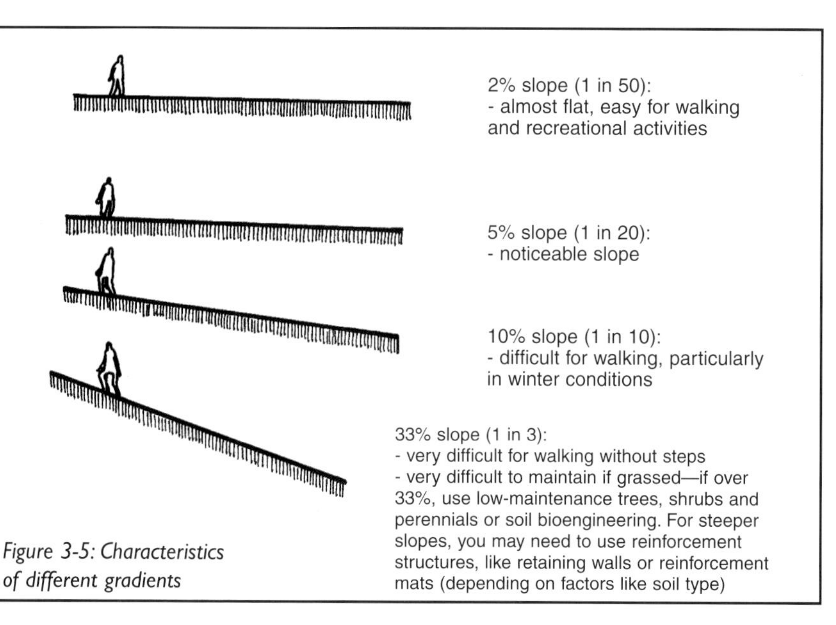

Figure 3-5: Characteristics of different gradients

your property once the grading is finished. It is preferable to keep and reuse the existing soil on your site, particularly the topsoil, to reduce costs, provide healthy soil for plants and reduce environmental impacts. Energy is used in transporting this material and using topsoil that does not come from your property removes an important natural resource from another property.

2. Balance Cut and Fill

The grading plan is also used to indicate the areas of cut and fill. Filling consists of adding soil and other materials like rock to raise the level to the desired elevations. Cutting is the excavation of a ground material to lower its finished level. The amount of material removed from certain parts of the site should be relocated elsewhere on the site itself in order to avoid bringing materials from the outside. Based on the proposed grading plan, the designer can calculate the quantities of soil to be cut and filled. Try to balance the quantities of materials cut and filled, to avoid the energy and costs associated with trucking them in or out.

When working with cut-and-fill situations in planting areas, it is important to avoid over-compaction of soils by machinery. If a subgrade is compacted and buried with fill, the compacted layer can be a barrier to water and air movement in the soil. This can create perched water tables, poor soil drainage, and can drown plant materials. To avoid these compacted layers, "scarify" or roughen the subgrade to remove surface compaction prior to adding fill. The objective of scarification is to provide a "rough transition" between layers of soil, rather than a compacted line.

Fill material
Existing grade
Cut material
Proposed grade

Figure 3-6: Balancing cut and fill

3. Preserve Existing Features

A new home should be placed so that as much of the original ground levels are preserved as possible, accounting for other objectives such as proper drainage around the house perimetre. Frequently in a residential development, fill is placed over tree roots, which can cause the tree to die, if the roots access to oxygen and water is insufficient. Shrubs and herbaceous plants are often completely removed and drainage is not the same as it was before development. All of these actions work against protection of the original site features such as trees. For more on protection of existing features, refer to the section on grade changes under trees in this chapter and to the section on preserving natural features and existing plants in Chapter 6.

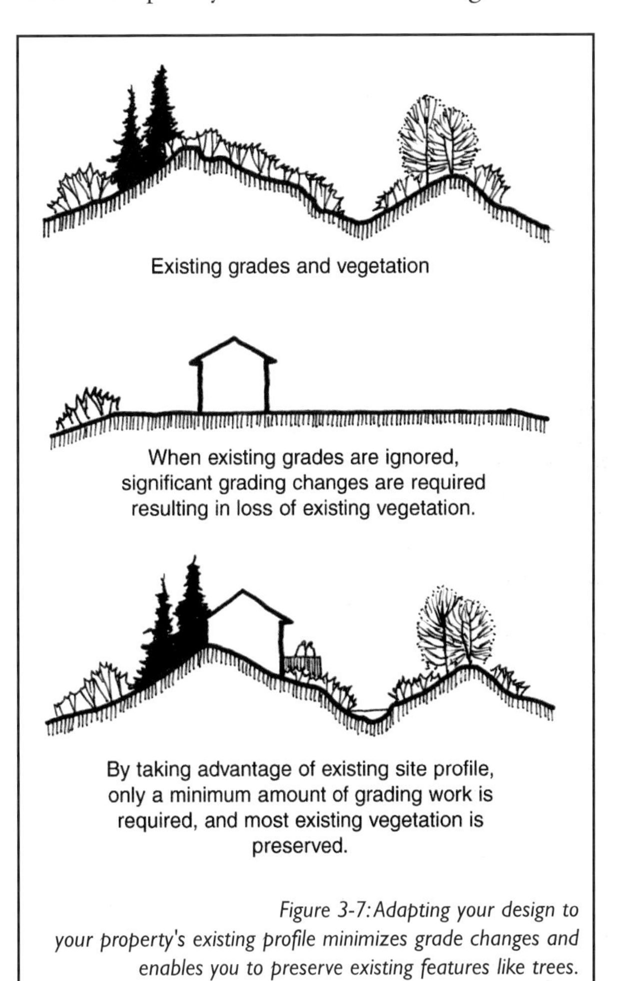

Existing grades and vegetation

When existing grades are ignored, significant grading changes are required resulting in loss of existing vegetation.

By taking advantage of existing site profile, only a minimum amount of grading work is required, and most existing vegetation is preserved.

Figure 3-7: Adapting your design to your property's existing profile minimizes grade changes and enables you to preserve existing features like trees.

4. Use Plants to Stabilize Slopes and Reduce Erosion

Plants help keep the soil in place on slopes and are very useful for slope stabilization and reducing erosion. Root systems keep soil particles together, while the upper part of the plant reduces the velocity of the surface runoff and the impact of raindrops that leads to erosion. Mulch is also useful. It covers the soil and prevents wind and water from dislodging soil particles.

Choosing the right plants for the slope is important, since there are different conditions depending on whether they are at the top, the sides or at the base of the slope. The top of the slope is often dry and windy. The sides, depending on their exposure and the slope, will also offer quite dry environments. They should be planted with deep-rooted vegetation that stabilizes the slope and that tolerates drier conditions than the vegetation planted at the base of the slope, which is usually moister.

Multi-layered, dense plantings, like trees and shrubs, provide more of these environmental benefits than shorter, single-layered ones, like lawns, and usually require less maintenance. Although initial costs are usually lower for lawns, long-term costs are usually higher due to higher maintenance needs. If you already have a naturally planted slope, take advantage of the benefits by leaving it alone, including fallen leaves, branches, logs and rocks that are already there.

Living by Water: If you have a slope beside a creek or other water body, look in the section on Living by Water at the end of Chapter 4 as well as the list of resources.

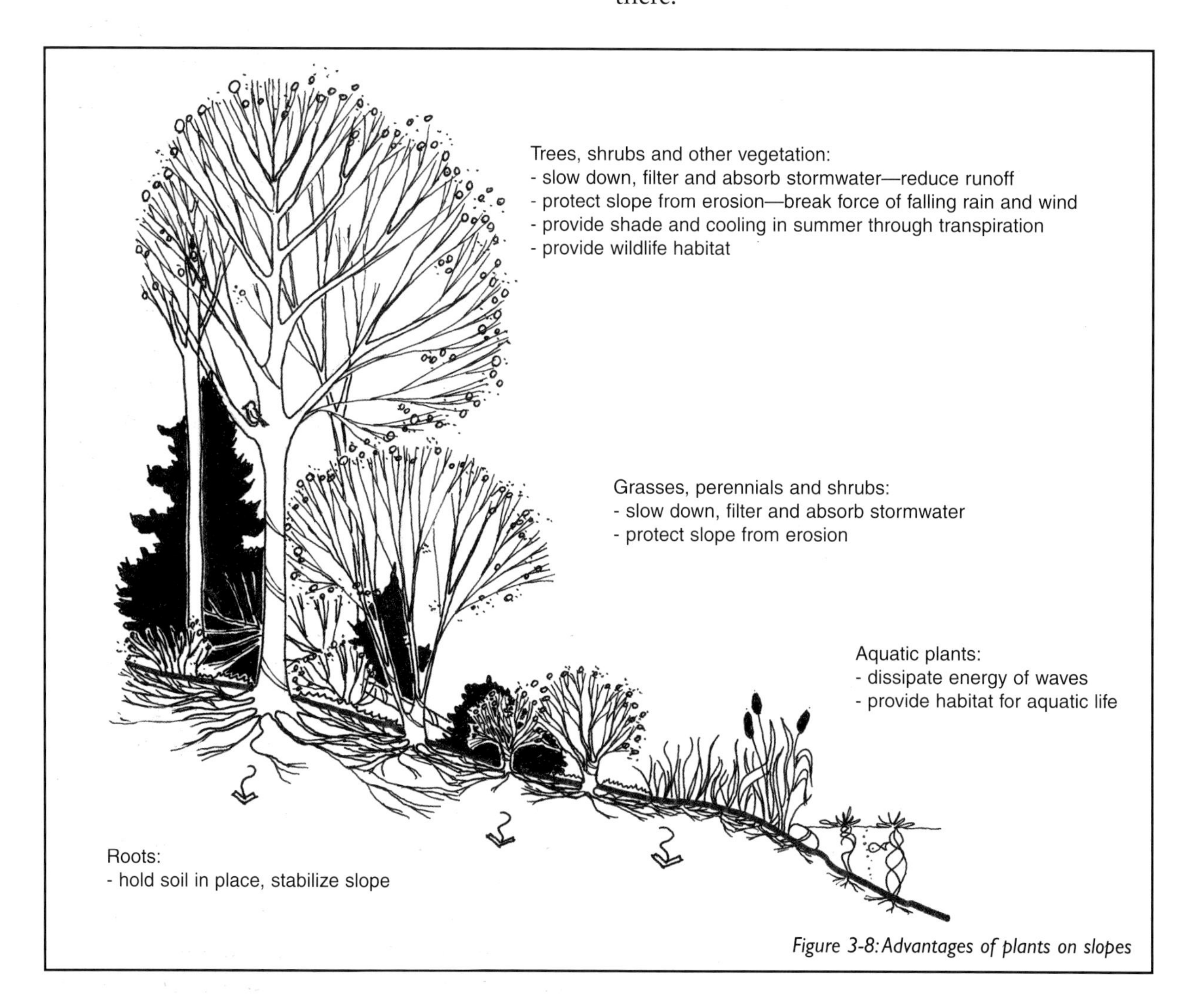

Figure 3-8: Advantages of plants on slopes

Advantages of plants on slopes

Plants make excellent slope stabilizers and control erosion. Their roots keep soil particles together, while their branches, stems and leaves slow down runoff and the impact of raindrops and wind that leads to erosion. They have the added advantage of letting stormwater soak into the ground, reducing runoff and improving water quality by trapping substances like soil particles. Plant communities that emulate natural environments, such as woodlands, also create wildlife habitat and provide a natural look to your property. Trees and other vegetation provide cooling and shade in summer and protection from wind in winter. In addition, planting techniques like soil bioengineering (see below) are self-repairing and are cheaper to install than hard structures like concrete retaining walls.

Figure 3-9: To restore habitat in this stream, several meander bends were reshaped and habitat structures were placed in the stream. Above, the crew puts down erosion control blanket using live stakes that will soon root and grow to re-vegetate the slope. This will help to stabilize the slope, provide shelter and shade, and control erosion.
Photo: Shawn R. Taylor, R.P. Bio: President, Habitat Works! Inc.

Trees, shrubs and groundcovers as slope stabilizers

Trees, shrubs and ground covers are excellent at stabilizing slopes and controlling erosion, since their interconnecting roots help to hold the soil in place. The leaves, stems and mulch help to absorb the impact of rainfall and reduce soil erosion caused by water and wind. Their roots are usually deeper than those of grass, so they can stabilize more effectively. Also, when the right plant is chosen, they can have long-term maintenance advantages over grass.

If you need rapid erosion control for large areas, species that spread via root-forming stems that grow horizontally above or below the surface are the most effective. Examples could include sumac, raspberries, snowberry, red osier dogwood and elderberry. For stabilizing slopes beside a waterbody, use plants on areas that are above water or only seasonally underwater.

Mulch helps protect the slope surface and reduce erosion. Erosion control blankets (see description below and figures 3-9 and 3-10) placed on the surface and staked into the ground offer a greater degree of slope protection and stability. Cut a notch in the blanket where trees or shrubs are planted. Refer to Chapter 6 and 7 for more information on plant design, installation and maintenance. Figure 3-10 shows how to put shrubs on a slope.

In some cases, you will need to combine plants with hard materials, such as logs or wooden cribs to stabilize steep slopes. There is a range of manufactured products available that are designed to stabilize slopes and that can be combined with plants. Contact a geotechnical engineer or supplier of erosion control products for more information. Soil bioengineering techniques can be used.

Lawn as a slope stabilizer

Lawn can be installed on slopes. It gives quick results but demands long-term maintenance, like mowing, which can be difficult on a slope, particularly slopes over 33 per cent. On a steeper slope, you'll need to anchor the sod strips with wooden pegs so they won't slide. Seeding the lawn on slopes is also an option. You can choose from a range of seed mixes adapted to slopes that require less maintenance and tolerate dry conditions. You can also choose slower-growing

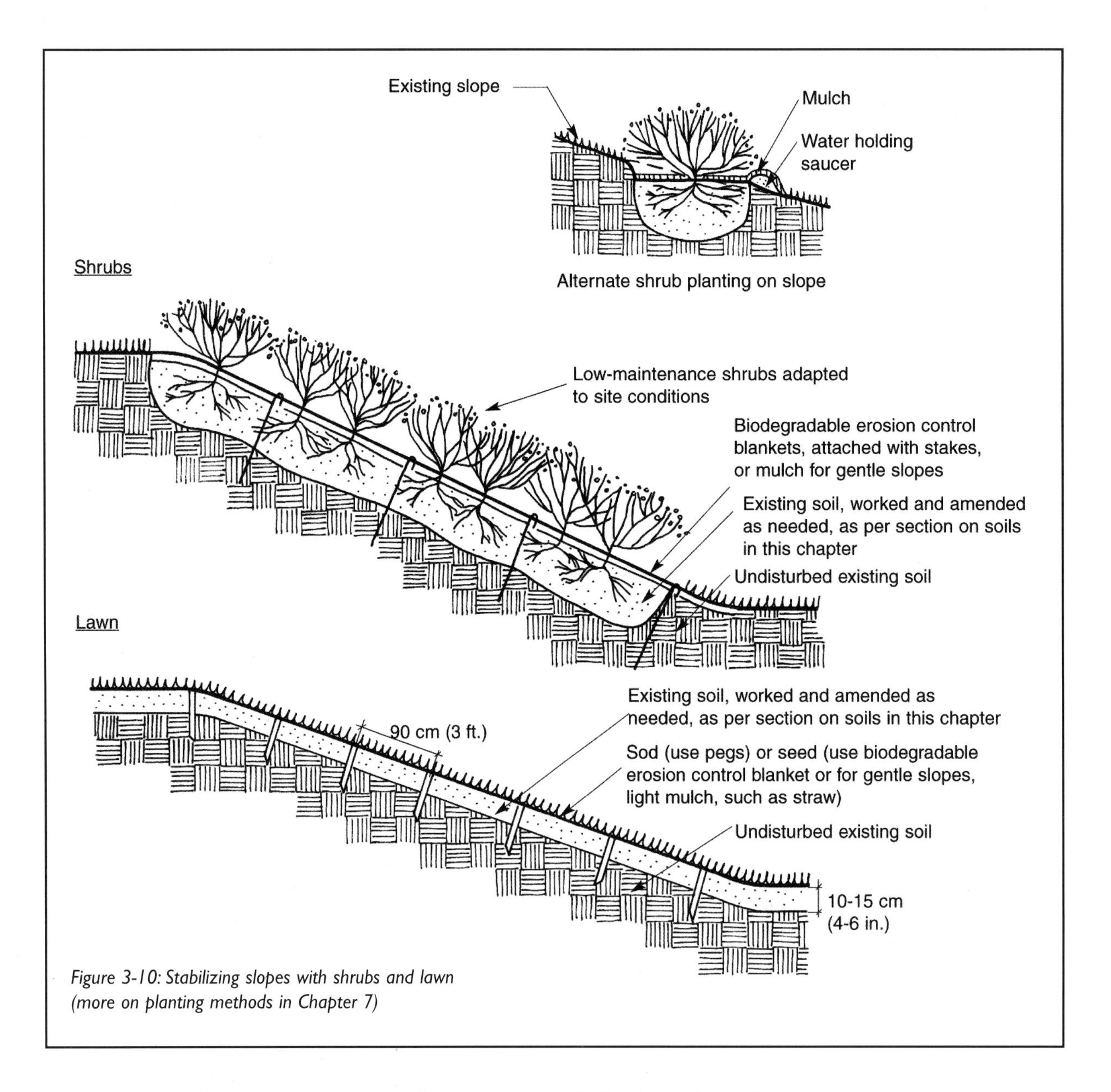

Figure 3-10: Stabilizing slopes with shrubs and lawn (more on planting methods in Chapter 7)

species, to cut down on mowing. Refer to Chapter 6 and 7 for more information on design, installation and maintenance of lawns.

When seeding on slopes, consider using an erosion control blanket after applying the seeds. They are generally made of a biodegradable mulch material, like straw or coconut fibres, held together with a light mesh, that either biodegrades or breaks down in the sun. Spread them over the soil and keep them in place with anchoring stakes. They prevent soil and seeds from washing away. Once established, the plants take over as the soil stabilizers. For large areas, you can hire a company to hydroseed, which involves spraying the area with a slurry of water, mulch and seeds.

Soil bioengineering

Soil bioengineering is a slope stabilization and erosion control technique that takes advantage of the ability of some plants, like willow and poplar (suitable species vary from region to region) to sprout from live woody cuttings, like branches, placed in the soil. Cut and plant them in the dormant season, such as spring or late fall. If they are not allowed to dry out, they will soon develop roots and leaves, and rapidly grow to vegetate and help stabilize the slope. This approach has several advantages including blending in with the natural landscape, being self-repairing and less expensive than hard structures. It is often used to protect banks along creeks and rivers.

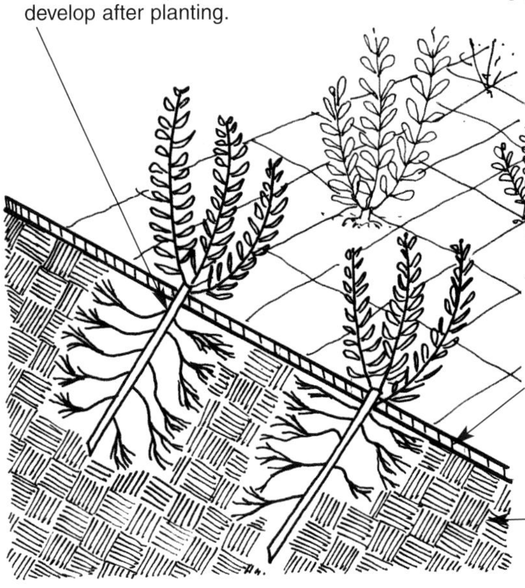

Live stakes: straight, dormant live cuttings of fast rooting woody plants—75-100 cm (2 ft. 5 in.- 3 ft. 3 in.) long, 2-5 cm (3/4 - 2 in.) diameter—tamped into soil with rubber mallet. Keep moist until installed. Roots, shoots and leaves are absent during planting, but develop after planting.

Cover with biodegradable erosion control blanket or for gentle slopes, use mulch.

Existing soil worked and amended as needed, as per section on soils earlier in this chapter

Figure 3-11: Live stakes - soil bioengineering

Live staking is one example of soil bioengineering (Figure 3-9 and 3-11). Live stakes are straight live cuttings of fast rooting woody plants, with no leaves or side branches, that are tamped into the slope with a rubber mallet so that they can take root in the ground and grow leaves and branches above ground. They are usually about 75-100 cm (2 ft. 5 in.-3 ft. 3 in.) long and 2-5 cm (¾-2 in.) in diameter. Ensure that cuttings do not dry out—you can store them in water before installation. You can cut the bottom at an angle to create a pointed end that is easier to insert into the ground. Spacing depends on soil conditions and species, usually about 30-90 cm (1-3 ft.) apart. The steeper the slope, the closer the row spacing. Make a pilot hole in the ground with a bar. Bury cuttings 3/4 of their length on wet sites and 7/8 of the length on drier sites, leaving the rest exposed. Mulch or use erosion control blanket between the live stakes.

For steeper slopes, you can combine plants with structural elements like crib walls made of logs, vegetated gabions (see description later in this chapter) and manufactured products designed to stabilize slopes. For example, geogrids are high tensile strength synthetic grid panels that are wrapped around layers of soil back into the slope. You can place rows of dormant live cut branches between the layers of wrapped soil. (Figure 3-12) Contact a landscape architect, geotechnical engineer or supplier of erosion control products for more information. If combining plants with these hard elements, ensure that the plants have appropriate soil medium to grow in.

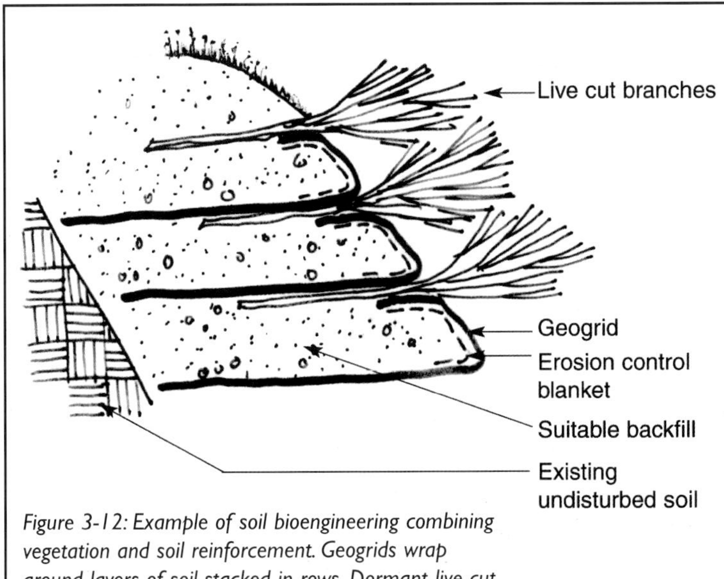

Figure 3-12: Example of soil bioengineering combining vegetation and soil reinforcement. Geogrids wrap around layers of soil stacked in rows. Dormant live cut branches (brush layering) from suitable species, for example, willows, are placed between rows.

The most suitable location for plants should be considered. For example, for slopes beside waterbodies like streambanks, trees and shrubs that prefer dry conditions are best on the upper parts of the slope, and water loving plants should be placed closer to the water's edge. Erodable areas that are permanently underwater should be protected with rock, whereas parts of the slope that are only seasonally flooded or above water can be stabilized with solutions that include plant materials.

Live cuttings must be kept cool and moist at all times before planting. If you are collecting many cuttings, collect them from over a wide area to avoid stressing any one plant and ensure genetic variety.

5. Retaining Walls: A Solution for Steep Slopes

When slopes are steep, structural reinforcement, such as retaining walls or reinforcement mats may be needed to stabilize them. The need for structural reinforcement depends on a variety of factors like the steepness of the slope, soil type and the vegetation cover. If you have any questions, contact a knowledgeable professional, like a geotechnical engineer, for advice.

This section focuses on landscape retaining walls, like low garden walls. The soil behind a wall can exert considerable pressure, so walls that are more than 1.2 m (4 ft.) tall or that are critical to the stability of your house foundation should be planned and approved by a professional engineer. Check with your local municipal building department. When the height between two levels exceeds 1.2 m (4 ft.) it is easier and more attractive to design several lower walls, if possible, to achieve the correct level.

Retaining wall foundations must also be laid on a stable base or on compacted crushed stone. A stable foundation will help prevent collapsing or movement of parts of the wall. You can enable drainage through the wall and eliminate the pressure caused by water or freezing behind the wall by installing crushed stone behind the wall, with perforated drainage pipe and weeping holes. Weeping holes in the wall at 1.5 m (5 ft.) intervals allow water to run through to the surface below the wall. There may also be different municipal regulations for the installation, type and height of walls and railings as well as drainage details. Check the local regulations.

Healthy retaining wall materials:

- More durable materials can be less costly in the long run, since they have lower maintenance and replacement costs.
- Transportation energy is reduced by choosing materials harvested and manufactured locally, for example stone that comes from your region rather than being transported from far away. Some materials may even be available from your own property, for example, stone and soil for filling gabions.
- Favour materials that are or can be reused and recycled.
- Consider whether the manufacturing process is polluting or energy-intensive.

Natural stone

There are two possibilities for building a stone retaining wall: dry stacked or stone and mortar walls, which are more costly but more stable. A mortared stone wall is made up of many stones affixed to each other with mortar. The resulting structure should be installed on a very stable soil or foundations below the frost line, otherwise movement can cause the joints to crack. To drain the surplus water, install a drainage pipe at the base of the wall and weeping holes in the wall itself. Mortar walls can sometimes degrade quickly due to disintegration of the mortar.

The dry-stacked stone retaining wall, built without using mortar or any other binding agent between the stones, is the most common. Small plants can be planted in the gaps between stones. It is preferable to use large stones with a minimum thickness of 7.5 cm (3 in.), 30 cm (1 ft.) depth and generally flat sides, to give the wall stability. Since they are held up only by their own weight, the maximum recommended height of a dry-stacked stone wall is 1.2 m (4 ft.). Just like the other walls, stone walls should be erected on stable foundations with adequate drainage behind the wall, for example by using crushed stone. Since the stones are

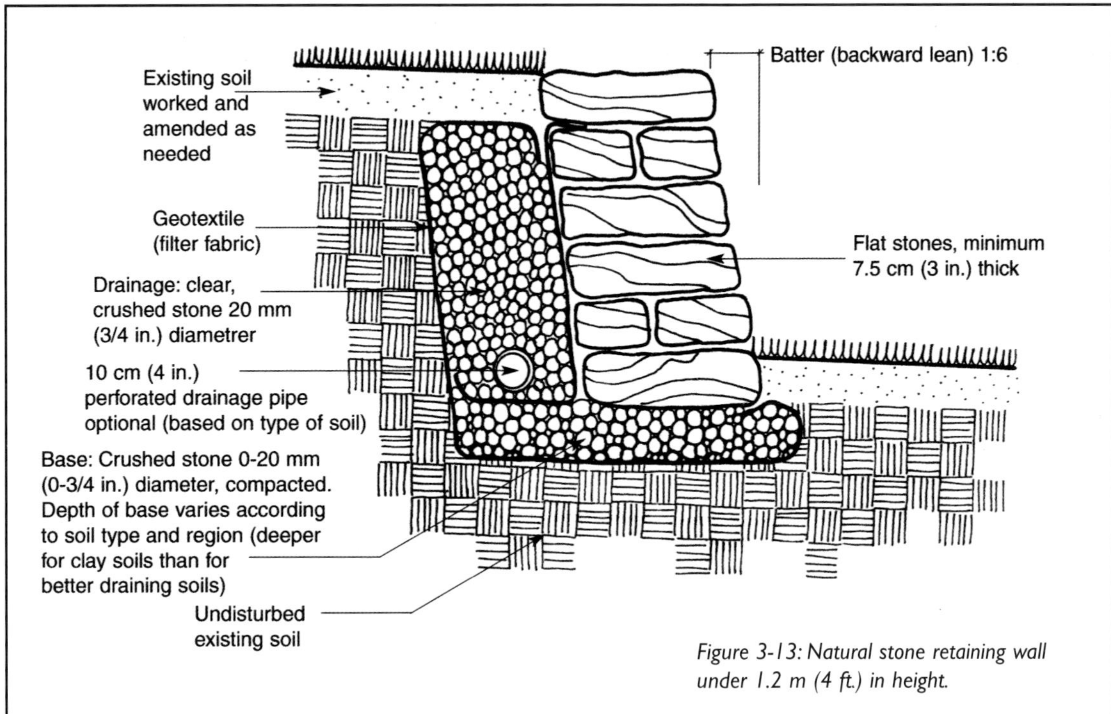

Figure 3-13: Natural stone retaining wall under 1.2 m (4 ft.) in height.

irregular, place a geotextile (filter cloth) behind the wall to allow drainage while preventing soil from infiltrating between the stones in the wall.

Figure 3-14: Dry stone retaining wall

Stones in dry-stacked walls are more reusable than mortar walls, since they are not mortared together. The lack of mortar also enables the stones to shift with the ground movement underneath them and be more easily repaired. Some weeding of undesirable plants between stones may be needed. Periodic resetting of shifted stones may be needed. Initial costs are relatively high, but so is the durability.

Building a dry-stacked natural stone wall:

- Group stones by thickness and alternate rows of various thicknesses, for example, one row of 7.5 cm (3 in.), one row of 10 cm (4 in.), one row of 7.5 cm (3 in.) and so forth.
- For stones that are not of equal thickness, use a stone chisel and a small mallet to shape them. Otherwise, use stone dust to level them.
- For stability, the face of the wall should not be vertical. The wall should lean back into the slope at an incline of 1:6 or 17 per cent. The backward slope is called a batter.
- Joints should alternate.

Gabions

Gabions are woven wire mesh baskets, filled with stones and connected to make a wall. They are flexible and permeable structures. Typically 1 m- (3 ft. 3 in.) high rows may be stacked, stepped back and tilted into the slope. They are used primarily on embankments because they are free-draining and can withstand the freeze/thaw action of water. Gabions are durable, medium-cost solution with nominal maintenance needs.

Soil can be added to them to enable plant growth. As described in the section on soil bioengineering, they can be used in combination with vegetation, for example, by placing rows of live cut branches (brush layering) between rows of gabions. When using vegetated gabions, fill them with a 50/50 mix of soil and stone above the normal water level, but the mixture should be about 70 per cent stone for parts of the slope at or below the water level. To contain the soil, use a biodegradable erosion control blanket inside the gabion. When combined with plantings, they are more attractive, absorb and slow stormwater, enhance habitat and provide other advantages.

Before

One year later

Figure 3-15: The extreme foreground of the top photo shows vegetated gabions with live stakes; the middle-ground shows brush layering with dormant live cut willow branches between the rows, just after installation. There is an erosion control blanket just behind the vegetated gabions and geogrids. The bottom photo shows growth after about a year.
Photo: Maccaferri Canada Ltd.

Poured concrete

Retaining walls are also often made of poured concrete. They involve more complex construction, and therefore it is best to get the help of a professional. They can be more costly than the other options. The footings for the concrete walls should penetrate below the maximum frost line and be installed on undisturbed soil or on compacted crushed stone. These walls should be equipped with weeping holes and crushed stone to allow drainage from behind the wall. If properly built, concrete retaining walls can be long-lasting and low maintenance, which can save you money and time later. However, if repairs are needed, they can be costly, since soil movement and wall damage are not isolated to individual blocks that act independently. One environmental disadvantage is that cement processing is polluting and energy-intensive.

Figure 3-16: Precast concrete block retaining wall

and provide more than one row of weeping holes.

Wood retaining walls over 1.2 m tall or that are critical to the stability of your house foundation should be pressure treated with a preservative to resist decay. For more information on the health and environmental impacts as well as basic precautions of using pressure treated wood, refer to Chapter 5. Retaining wall alternatives to chemically treated wood are listed above.

Precast concrete blocks

Precast concrete retaining walls are less expensive than concrete walls poured on site, but just as durable. Because these walls are made of many independent blocks, they can handle slight soil movement without affecting the structure and repairs are possible since the blocks can be removed and replaced. Removed blocks are reusable.

Now available in a vast range of textures, shapes and colours they can be used in various designs. Refer to manufacturers' information for installation details. You may also be able to find a system that enables gaps in which vegetation can be planted. Plants enhance the wall's appearance, absorb and slow stormwater, and enhance habitat among other advantages. Again, cement processing is polluting and energy-intensive.

Wood

Wood is a renewable material that is easily obtained in Canada and economical to buy. It is subject to decay and deterioration, particularly when used in-ground. Its longevity can be improved by providing good drainage, for example by installing a 30 cm (12 in.) wide layer of crushed rock behind the wall along its full height, surrounded by a filter fabric. Also install a perforated drainage pipe in the crushed rock at the base of the wall and outlet it at both ends,

6. Avoid Grade Changes Around Trees

Changing the existing grade around trees can result in serious decline or even death of the tree. Raising the grade prevents air, moisture

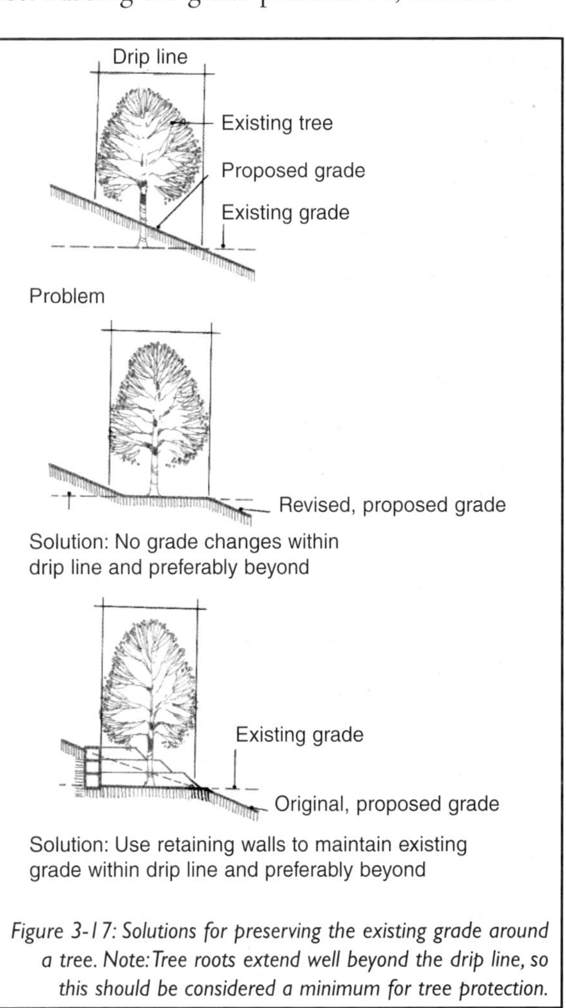

Figure 3-17: Solutions for preserving the existing grade around a tree. Note: Tree roots extend well beyond the drip line, so this should be considered a minimum for tree protection.

Figure 3-18: These trees are at risk of dying because the bulldozer has moved soil over the roots and driven over them repeatedly, resulting in soil compaction and difficulty for the roots to access water, nutrients and air. Proper planning and management as well as protective fencing, like in Figure 6-3, would have prevented the grade changes and access by this heavy equipment.

and nutrients from reaching the roots. Retaining walls can be used to maintain the existing grade around trees when there are grading changes near them. (Figure 3-17)

During your site inventory and analysis, you identified the trees you want to preserve. The next step is to consider how the trees will be protected during and after construction. Consult a professional, for example, a landscape architect or certified arborist, to evaluate the trees you want to preserve, and to propose tree protection methods before work begins. Refer to the section on tree preservation in Chapter 6 for more details.

For best results, the protective zone should extend beyond the dripline of the tree (the

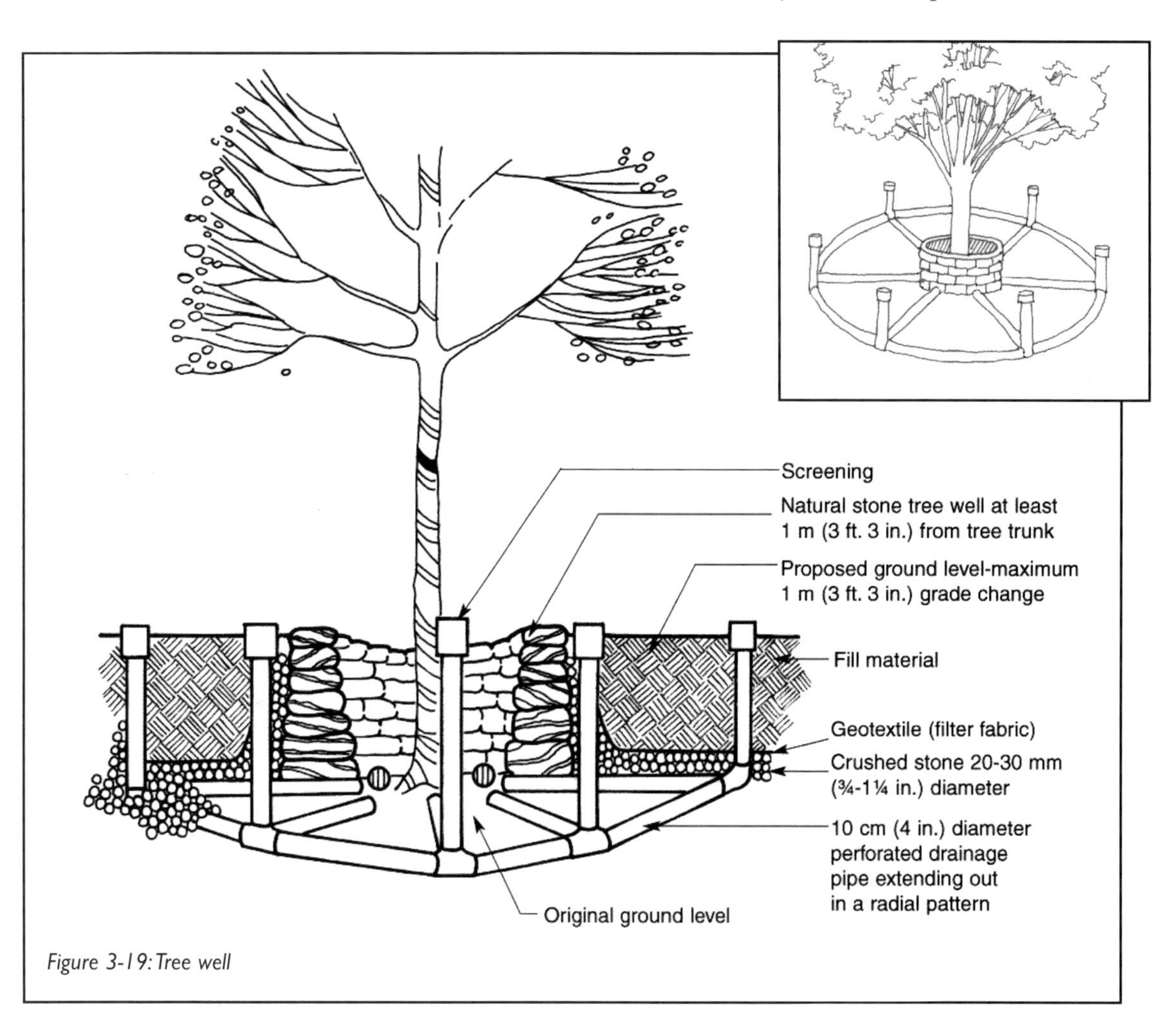

Figure 3-19: Tree well

ends of the outer branches) as tree roots normally spread out beyond the dripline. No grade changes, filling or other construction-related activities should be done within this area. This area contains most of the roots that take up essentials like water, nutrients and air.

During construction, you should fence the area to protect it from heavy equipment and compaction from storage of heavy materials as well as grade changes. (Figure 3-18)

It is possible to raise the grade over the root area of a tree you want to preserve, by building a tree well around it with a system of perforated drainage pipes to help the roots access water and air (Figure 3-19). Place the pipes on the original grade radiating out from the tree well extending at least to the dripline of the tree and preferably beyond, covered by a layer of crushed stone, which is covered by a geotextile (filter cloth), then with fill material. Vertical perforated drainage pipes from the surface to the crushed stone layer will help bring water and air to the roots. Tree wells can be expensive. Before building them, analyze the health, size and species of trees, soil conditions, drainage patterns and depth of fill.

For more information on tree preservation, refer to Chapter 6.

Do you live in an area of Canada where sensitive clay soils can be found, like Ottawa or Montréal? If you have noticed damage to your foundation and are concerned about the causes, refer to the CMHC *About Your House* fact sheet *Understanding and Dealing With Interactions Between Trees, Sensitive Clay Soils and Foundations.*

Chapter 4

STORMWATER AND WATER CONSERVATION

MANAGING STORMWATER

Stormwater is rain or melted snow and ice. Ever wonder what happens to water that falls on your roof and driveway? Typically, it is directed quickly toward the street and into the municipal sewer system. Usually, the stormwater runoff ends up in lakes, streams and other watercourses with adverse affects to aquatic habitat and water quality, for example through erosion and the introduction of oil, road salt and other pollutants. This chapter provides tips on managing stormwater on your property to help reduce these impacts. It also provides other water-related landscape tips, such as water-efficient irrigation and water ponds.

Did you know that municipal water consumption doubles in summer, due to outdoor activities like lawn and garden watering? One of the best ways to save water is to select plants that are suited to your rainfall and site conditions. Irrigation systems will be discussed in Chapter 4. In Chapter 6, we will talk about xeriscapes and other planting approaches that save water. Chapter 7 also discusses water saving plant maintenance techniques.

Let Nature Inspire You

In nature, when a drop of water falls to the ground and soaks slowly into the soil, it will make a long journey before reuniting with a watercourse or the water table. Rainfall that soaks into the ground may be taken up by a plant that will then give it off in the form of transpiration, causing water vapour to rise into the air. Evaporation also returns water to the air. It then condenses and returns to the ground again in the form of precipitation. Water is used and reused in an endless cycle known as the hydrologic cycle. (Figure 4-1)

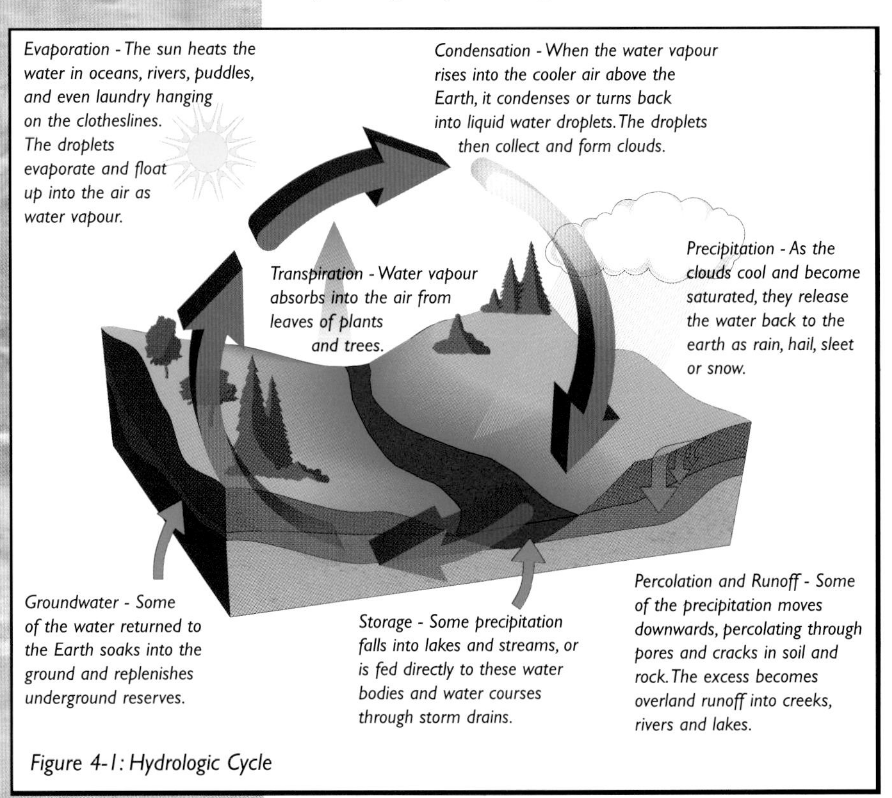

Figure 4-1: Hydrologic Cycle

In contrast to water's natural cycle, stormwater runoff in urban areas is usually directed from roofs and pavements to storm sewers and ditches (Figure 4-2). Instead of soaking slowly into the ground, the water is flushed quickly off

Figure 4-2: The fish pattern on this storm sewer cover is a reminder that substances that drain into the sewer end up in water bodies like rivers and lakes, affecting water quality and aquatic habitat.

the property, picking up substances like de-icing salt and oil from roads as well as pesticides and fertilizer. Storm sewers often outfall to local streams. The increased volume and frequency of high flows in the stream causes erosion and sedimentation. In cities with combined storm and sanitary sewers, the system can become overloaded during heavy rainfalls or spring thaw, and overflow from the sewer system to natural waterbodies before treatment. All of these changes can dramatically impact aquatic habitat and water quality in our lakes, rivers and streams, and at times leads to beach closures.

Drawing on nature's system, the water could be absorbed into the soil, where it would be filtered. It would then soak slowly into the water table and be used to sustain plants.

Tips for Managing Stormwater on Your Property

When doing your grading plan (see Chapter 3) think about how you will deal with stormwater. The proximity of the neighbours, particularly on small lots, and municipal bylaws may require you to drain surface water toward the street where it is conveyed into the municipal stormwater system. Some local governments may require you to capture a certain percentage of rainfall on your property, for example by letting it soak into the ground on your property, which offers many benefits.

Tips on stormwater infiltration are provided in this chapter. Your approach to stormwater management on your property can combine several of the approaches described in this chapter. You will want to drain stormwater away from some locations, like away from the house, and let it soak into the ground in other locations, for example by using a swale directing the water toward a rain garden. To verify if your overall stormwater approach complies with regulations, consult your municipality.

Here are three principles to consider:

1. Drain water away from your house. Avoid drainage problems to neighbouring properties.
2. Let stormwater soak into the ground on your property.
3. Recover and reuse stormwater.

1. Drain water away from your house. Avoid drainage problems to neighbouring properties and key activity areas.

Keeping runoff away from your house can prevent water problems in your basement or foundation. Follow the tips below, to avoid foundation drainage problems in your house. (Figures 4-3 and 4-4)

Ground surface at the house should be at least 15 cm (6 in.) below the lowest course of bricks or 20 cm (8 in.) below stucco or siding

Slope for drainage away from the house at least 5 per cent for 2 m (6 ft. 6 in.) and 2 per cent for pavements adjacent to the house.

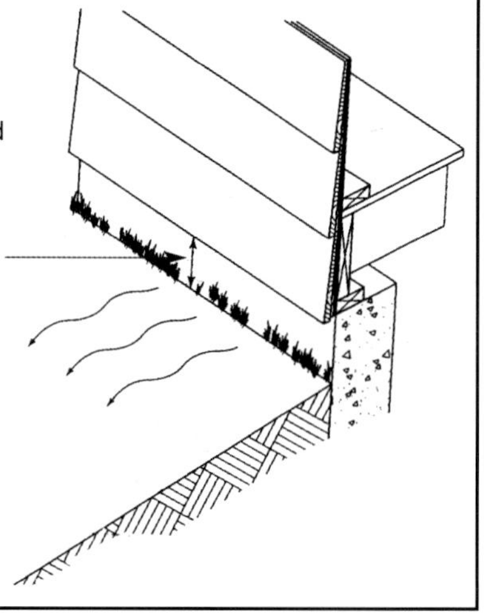

Figure 4-3: Slope away from the house

Figure 4-4: There are three potential sources of moisture problems for this house. The soil in this raised bed is too high up the brick surface of the house. The ground surface at the house should be at least 15 cm (6 in.) below the lowest course of bricks. Also, the grade of the soil in the raised plant bed is sloping toward the house and the downspout ends abruptly right at the wall. These conditions will drain water directly to the house wall and foundation, potentially resulting in moisture problems.

Planted swales and French drains are among the options you can consider for conveying stormwater from one location to another. These and other conveyance methods such as ditches, catch basins and drain inlets are discussed later in this chapter.

Drain stormwater from your main activity areas, such as patios, driveways and sheds, where you don't want water to accumulate. Generally, at least a 1 or 2 per cent slope is needed where you want drainage for lawn areas and 1 per cent for patios, decks

Keeping moisture away from your house:

- Slope the ground away from your house around the entire perimeter. While at least a 5 per cent slope away from foundation walls is acceptable, a 10 per cent slope for the first 2 metres is preferred to ensure the correct slope after the soil settles and compacts. Use at least a 2 per cent slope for impermeable surfaces next to the house, such as a driveway or patio. Check and adhere to local municipal standards.
- The ground surface at the house should be at least 15 cm (6 in.) below the lowest course of bricks and at least 20 cm (8 in.) below the lowest level of non-masonry finish, like stucco or siding.
- If you are planting adjacent to the house, select species that are suited to local rainfall conditions, so that minimal irrigation is needed. If you do need to water your plants, avoid wetting the building.
- If you are planting adjacent to the house, you can install an impermeable membrane, such as a layer of clay or sheet of polyvinyl, under the soil, allowing enough soil depth to accommodate the roots of plants above it. Run the membrane along the foundation and along the ground at least 1.2 m (4 ft.) sloping away from the house to help drain water from the foundation.
- Place an extension on your eavestrough downspouts to direct roof runoff away from the house. Run the extension toward a rain garden or soakaway pit (both of these features are described later in this chapter) at a safe distance from your house, usually at least 4 m (13 ft.). Use a splash pad or loose, hard material, like pebbles or crushed rock, at the end of the extension to absorb the impact of the water on the soil. You can bury the extension to make it less intrusive, but clogging and other failures will be more difficult to observe than for above-ground downspout extensions.
- For more details, including roof drains and foundation drains refer to the CMHC publication *Investigating, Diagnosing and Treating Your Damp Basement.*

and walkways (depends on the permeability of the material). For slopes around the house, refer to the box above. Figure 3-4 will help you measure the slopes. As described later in this chapter, there may be some areas where drainage isn't needed, that can be relatively flat, like a lawn area at 0.5 per cent.

As a homeowner, you are responsible for the proper drainage of surface water on your property. A neighbour who is affected by a change in the drainage pattern on your lot has the right to ask you to correct the situation. Check your municipality for local requirements about drainage from your property to neighbouring properties. If a grading and drainage plan of your property seems too complex, it is preferable to call a professional, for example, a landscape architect or engineer.

Stormwater that runs along bare soil and slopes can result in erosion of the soil and sediments being sent to rivers and other waterbodies. Not only is a valuable soil resource lost, but water quality and aquatic habitat are also negatively affected. To prevent this, stabilize slopes on your property and ensure that bare soil is covered, preferably with plants. Refer to Chapter 3 for more details on establishing vegetation on slopes. There are also some tips at the end of this chapter on river banks and other slopes at the edge of water.

2. Let Water Soak Into the Ground

To reduce stormwater impacts, you can create areas where the water can soak into the ground where it won't interfere with other features. Consider these options:

a) permeable surfaces
b) rain gardens
c) flat areas
d) soakaway pits
e) soil aeration

a) Keep impermeable surfaces to a minimum

Impermeable surfaces, like regular asphalt driveways or concrete patios, prevent water from soaking into the ground. To reduce runoff from your property, limit the size of impermeable roof and paved areas and use permeable surface materials that allow water to soak into the ground. Analyze how much paved surface you really need for your regular activities and try to keep these surfaces to a minimum. By limiting the size of driveways and other pavements, you allow more space for planted areas, which is excellent for letting water soak in. For example, on driveways, you can install two strips of paving for vehicles and plant grass or a ground cover in the spaces between the strips (see figure 5-2).

Use permeable paving surfaces that will allow water to soak into the soil. This includes precast concrete pavers designed for infiltration or loose aggregate materials, such as crushed stone, pebbles or crushed brick. There are more details on permeable hard surfaces in Chapter 5. Grading pavements so that they drain towards absorbent ground will also help reduce and slow down runoff.

Roofs are also impervious surfaces. If you are building or expanding, consider compact structures of two or more storeys with a smaller roof area than a more spread-out single storey design. Green roofs with lightweight growing medium that holds water, and drought tolerant plants, are popular in Europe and their use is growing in Canada. Refer to the next chapter for more information on green roofs.

b) Direct stormwater toward a rain garden

A rain garden is a strategically located shallow depression that intercepts runoff. It can be covered with mulch and vegetation, such as perennials, shrubs and grasses that tolerate moist to dry conditions and/or loose, hard materials such as pebbles and river stones. Stormwater is directed to rain gardens to allow it to soak into the ground (Figure 4-5).

Figure 4-5: The owners of this property in Cortise, Ontario turned their backyard swale into a dry river bed rain garden. The bottom is covered with river stone and the sides are covered with woodchip mulch and tall grasses, perennials and shrubs. Photo: Glen Pleasance

Locate them at least 4 m (13 ft.) away from your house, in a natural low point that won't interfere with your main activities. Rain gardens should also be located in areas with course soil, like sandy to loam soil or any soil that is loam or courser. The bottom should be at least 1 m (3 ft. 3 in.) above the seasonally high groundwater table. Soils that are fine-grained, like clay soils or anything finer than a loam, can slow down infiltration rates and may result in longer-than-desired periods of standing water (maximum two days) or overflow. Compaction also decreases the infiltration rate of soils, by creating a "crust" at the surface that slows the rate of water penetration into the soil. Adding organic matter such as compost or rotting leaves to the soil increases the population of earthworms and micro-organisms, which keep the soil surfaces open. You should have suitable soil to a depth of 0.6-1.2 m (2-4 ft.). If these conditions are not present, it may be possible to amend your existing soil to this depth with something coarser grained and with organic matter, loosening the soil in the hole bottom and sides. The more shallow the depth of permeable soil, the more frequently the rain garden will fill and overflow.

Rain gardens usually have a curvilinear shape. You can use a garden hose to layout the shape on the ground. They should slope gradually down toward a low point about 7.5 cm (3 in.) deep near the centre for soils with low infiltration rates, like loam, and 15 cm (6 in.) for soils with high infiltration rates, like sandy or gravelly soils.

For information on removing lawn, refer to the section on bed preparation in Chapter 7. This chapter also provides planting tips. To keep it looking neat, leave a crisp edge, for example, with natural stone (Figure 4-6).

Figure 4-6: Rain garden

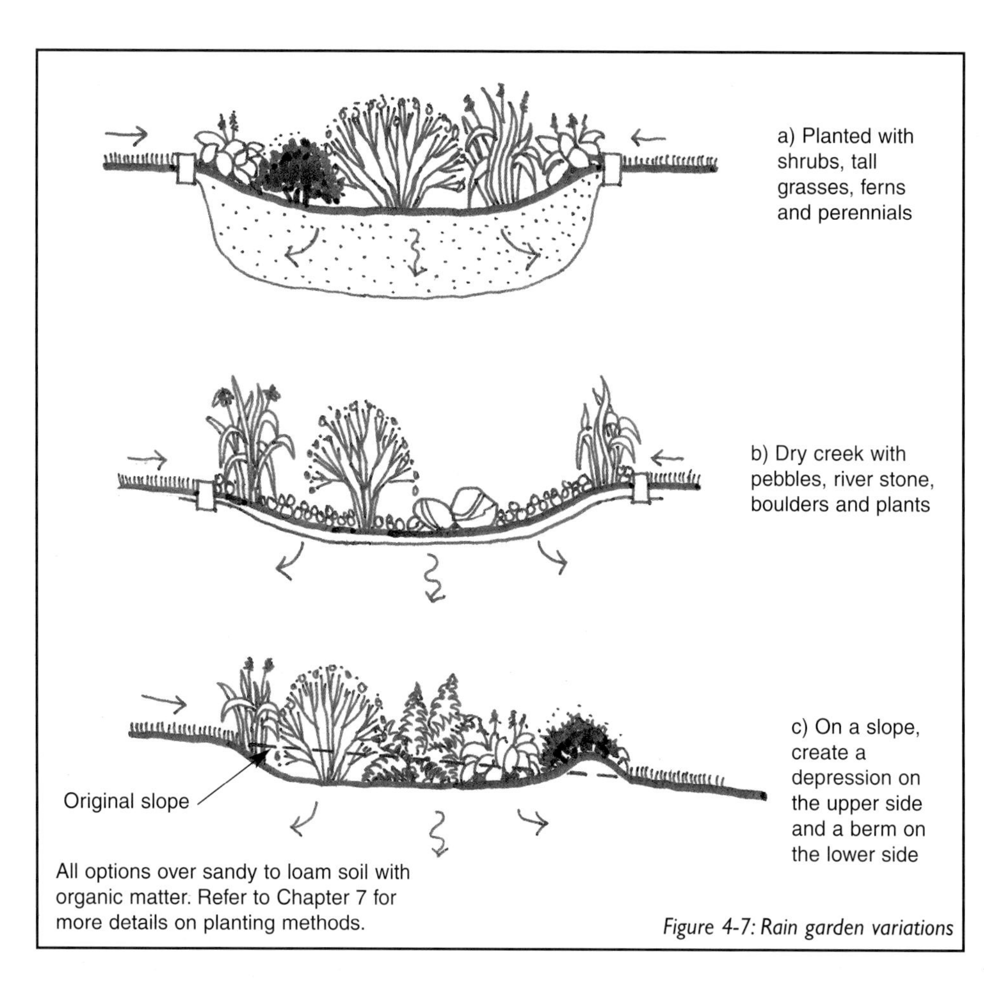

Figure 4-7: Rain garden variations

After a heavy rain, this area may not be useable for foot traffic and therefore should be located in a less frequently used spot, depending on the size of the depression and the area draining into it, as well as the type of soil and depth of water table. It is important to analyze your property to determine if it has the right conditions for a rain garden.

Figure 4-8: You can direct a roof downspout extension into your rain garden to receive roof runoff. To reduce the impact of the water flow on the soil, use a splash pad or pebbled area under the end of the downspout extension. Or it can discharge into a lawn area that flows into the rain garden.

It is important to note that rain gardens should not generally contain standing water. Stormwater directed to them should infiltrate into the soil and you should ensure that you do not have standing water for more than two days.

Select the best option (Figure 4-7) for your design and soil type:

a) Rain gardens can be covered with plants, such as perennials, shrubs, grasses or ferns that tolerate wet conditions. They also need to tolerate dry conditions that may occur during periods of low rainfall. Therefore select plants that are adaptable to a range of moisture conditions. Place plants that can tolerate fluctuating moisture conditions in the middle and deepest parts.

b) They can also be surfaced with loose, hard materials like river stone and pebbles. Water readily percolates through these materials, so the surface is generally kept dry. Use a geotextile (filter cloth) to keep soil separate from the pebbles and stones while letting water soak through. Larger stones and plants are placed in these areas to enhance their appearance. With a long and thin layout, they can look like dry creeks. (Figure 4-5) You can also place plants in this type of rain garden, particularly along the edge.

How to determine the size of a rain garden:

How much water flows in?

1. Estimate the area in square metres of the roof and impervious paved surfaces draining into the rain garden.
2. Estimate a target for how much rainfall should be captured in your rain garden over a 24-hour period. You can get rainfall data for your region from Environment Canada. Some municipalities set targets, so check with your local municipality. If no rainfall capture target is available, you can set one yourself based on your local rainfall conditions. An example is 25 mm of rainfall received over 24 hours. (This will vary from region to region, for example, it may be as low as 5 mm in some locations.)
3. To estimate the amount that will flow in, multiply the impervious area in square meters (step 1) times the rainfall capture target in metres (step 2). For example, a suburban home may have 150 m^2 of impervious roof and paving and a rainfall capture target of 25 mm over 24 hours. This leads to a need for 150 x 0.025 = 3.75 m^3 of water to infiltrate over 24 hours.

How much water flows out?

An analysis of the permeability of your soil is critical to the success of your rain garden. Determine the percolation rate of soil at the bottom of the rain garden. As a rule of thumb sandy soils have a minimum rate of 210 mm/hour or more. In sandy loam soil, a minimum rate would be 25 mm/hour, and for loam, the rate is 15 mm/h. In clay soils, rates can be as low as 1 mm/hour. Your local or regional government may have more information on conducting percolation tests.

What is the size of the rain garden?

If the percolation rate for a loam soil is 15 mm/hour, it will absorb about 36 cm of water per 24 hours. Convert the figure to metres (e.g. 0.36 m). Now, divide the estimated in-flow (from step 3) by the percolation rate. Using the example of a 3.75 m^3 target for infiltration, you'll need an area of 3.75 ÷ 0.36 = 10.4 m^2. Thus a rain garden of 4 m x 2.6 m (13 x 8½ ft.) would suffice. As described earlier, the permeable soil under them should be 0.6 m-1.2 m (2-4 ft.) deep.

c) If you are creating a rain garden on a slope, you'll need to dig down at the upper side of the slope so that the outer edges of the rain garden are level. Build a low berm along the lower edge of the slope. The berm can be built up using soil you removed at the higher end of the slope. Compact the berm and cover it with plants.

Overflow

Rain gardens are designed to infiltrate water from common rainfall events. However, more intense events can come along once or more per year, with extreme events every 25 or 100 years. To avoid flooding outside your rain garden during the more intense events, you may need to include an overflow system, depending on local requirements. This could be as simple as an in-ground perforated pipe or shallow swale that drains toward a part of the front yard that slopes toward the street. You could reduce the frequency of overflow by increasing the size of the rain garden area or depth of permeable soil.

If you already have a wetland or seasonal wet spot in a suitable place on your property, you can benefit from this natural rain garden by leaving it in its natural state.

c) Keep some areas relatively flat

Certain areas like the perimeter of the house and other areas where you don't want water to accumulate, like driveways, should be sloped so that surface water drains from them, as previously discussed. But other areas where drainage is not needed can be relatively flat to allow stormwater to soak into the ground and reduce runoff. For example, an infrequently used lawn away from the house could have a 0.5 per cent grade.

d) Let water soak in under the soil surface with a soakaway pit

Soakaway pits are underground pits for runoff that slowly percolates into the surrounding soil. The pit is filled with clear crushed stone surrounded by a geotextile (filter cloth). The residual spaces between the stones can accommodate a certain volume of water and let it soak into the ground.

To receive drainage from your rooftops, you can place the end of the roof downspout or leader directly in the pit 7.5-15 cm (3-6 in.) below the top of the pit along the full length of the pit. The portion of the leader extension in the pit should be perforated to allow water to flow out along its length. The soil cover over the pit depends on the soil characteristics, the frost heave potential and the depth of the bottom of the pit. For example, in Ontario, a 1 m (3 ft. 3 in.) deep pit in sandy soil should have about 65 cm (2 ft. 1½ in.) soil cover. Less soil cover would be needed in regions with less freezing. Shallow pits less than 1.5 m (5 ft.) deep are generally preferred. The maximum depth of the pit can be estimated by multiplying the soil percolation rate in metres/hour by the time you want the water to draw down, preferably 24 hours. For example, if loamy sand soil has a minimum percolation rate of 0.06 m/h and you want it to draw down in 24 hours, the pit should be no deeper than 1.4 m (4½ ft.). Refer to "how to size a rain garden" earlier in this section for percolation rates for other soil types.

The pit should be large enough to accommodate the targeted volume of runoff from the roof. Determine the volume by estimating the area of roof draining into it and the rainfall capture target as discussed earlier in "how to size a rain garden." This will give you a certain volume of water. When sizing the pit, factor in the space that the crushed rock takes up. Clear crushed stone in the 38 mm (1½ in.) size range will store about 33 per cent of its volume as water, while 5 cm (2 in.) diameter stone will store more.

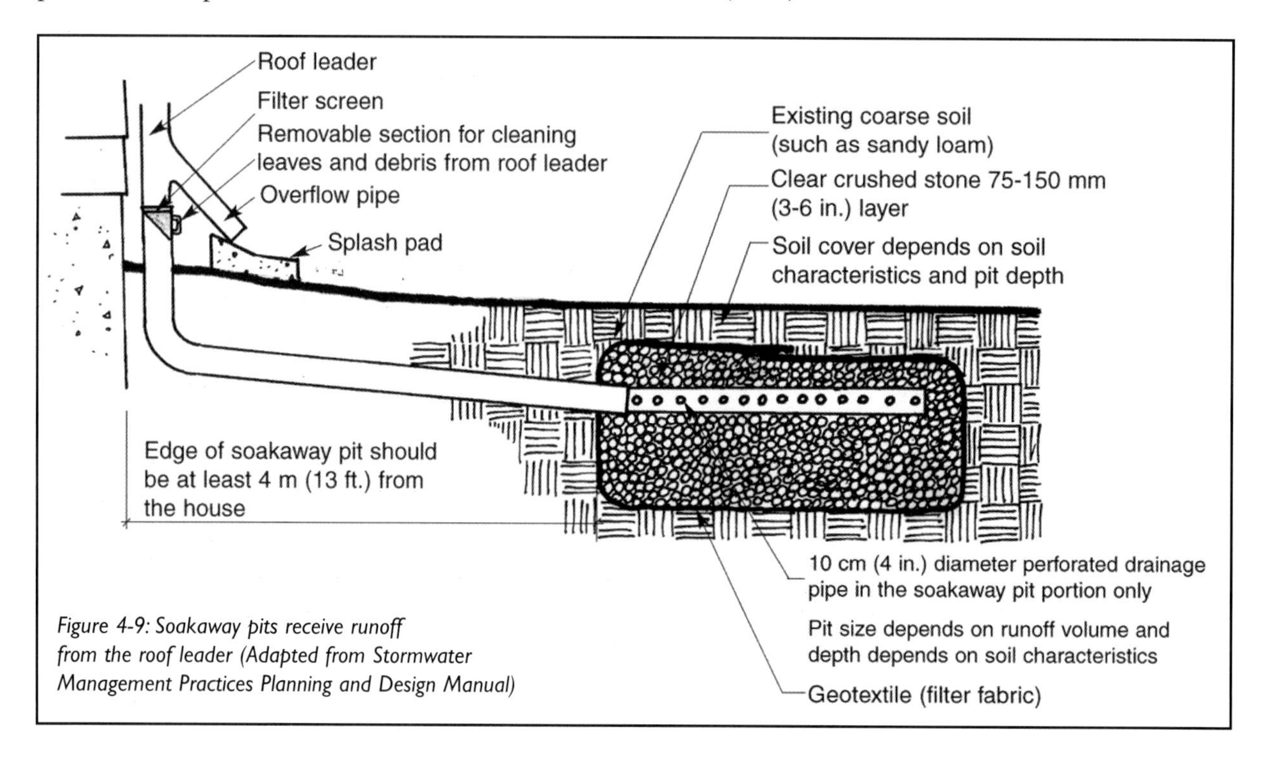

Figure 4-9: Soakaway pits receive runoff from the roof leader (Adapted from Stormwater Management Practices Planning and Design Manual)

When locating rain gardens and soakaway pits, select an area with uncompacted, permeable soil with a percolation rate of least 15 mm/hour (generally includes soil that is loam or coarser, like sandy loam) and where the water table is not too high, usually at least 1 m (3 ft. 3 in.) below the bottom of these features. Also the depth from the bottom of these features to bedrock should be at least 1 m (3 ft. 3 in.). The minimum distance to the house should be 4 m (13 ft.). The size of the contributing drainage area and runoff volume should also be factored in. It is important to analyze your property conditions to determine if you have the right conditions for these features.

For example, if the roof area draining into the pit is 50 m^2 and the rainfall capture target is 20 mm/24 hours, the pit should be sized to receive 1 m^3 of water. If you have 38 mm (1½ in.) crushed stone, the pit size should be about 3 m^3.

A removable screen should be installed on the roof leader above the ground where you can access it for maintenance. A filter on your roof leader will help trap leaves and other debris, and keep them from clogging your soakaway pit. Inspect and clean the filters in the autumn after the leaves have fallen. Check to see if water is frequently coming from the overflow pipe to determine if you need to clean or unclog the soakaway pit or roof leader screen.

The length of the pit (in the direction of inflow) should be longer than the width to distribute the water more evenly in the pit.

e) Aerate soil to reduce compaction

Soil compaction on lawns or other planted areas occurs most commonly on clay soils or in high traffic areas. Signs of compaction include water pooling for extended periods or flowing rapidly over the surface. You can reduce compaction by aerating your lawn, which is the removal of small plugs of soil, as described further in chapter 7. You can also gently break up soil in planted areas and amend it with compost.

Figure 4-10: The downspout directs roof runoff to this rain barrel. The tap at the base of the rain barrel enables collected water to be easily used in the garden via a bucket or hose, and overflow from the top can be directed to the garden. Keep it covered with a mesh screen to keep out insects.

3. Capture and Reuse Water

Your plants can take advantage of the stormwater passing through your property.

a) Direct stormwater to planting beds

You can grade your property so stormwater is directed to planted areas. You can also direct water from your roof downspout or leader to planting beds away from your house, using an extension. Refer to the section on rain gardens for more information.

b) Install a rain barrel under downspouts

By installing a rain barrel at the base of your eavestrough downspout, you can recover water from the roof runoff and build up a supply for watering your plants. An overflow at the top of the barrel with a hose directed to planting areas will send the overflow where it's useful rather than to your foundation. Also ensure that it is tightly covered with a lid or mesh screen to keep out insects. When the barrel is elevated above ground level, water can be distributed by gravity, using a tap and hose at the base of the rain barrel.

Water-efficient landscape: Ostrowski residence, Calgary

Jorg and Helen Ostrowski's home in Calgary incorporates a variety of water-efficient landscape features that are perfectly suited to this dry climate. This landscape thrives on whatever stormwater falls on the site without using a drop from the city water system.

Compost bins from reused wooden pallets and recycled plastic (1 and 2) are used to create organic soil amendments that improve soil moisture and nutrients in the vegetable garden (3) and fruit/ berry patch (5). Their pond (4), is fed by water from the cistern overflow. All pathways in the backyard (6) are made of materials that allow water percolation, like wood chip mulch, pebbles and natural stone. Under the ground are 15,000 litre (4,000 gallon) cisterns (7) that collect rainwater for use in the home. The deck and benches (8) are made from recycled plastic lumber. Dark, reused plastic pails on the south side of the house are a ready source of prewarmed water for watering plants. Water is also collected in rainbarrels (9).

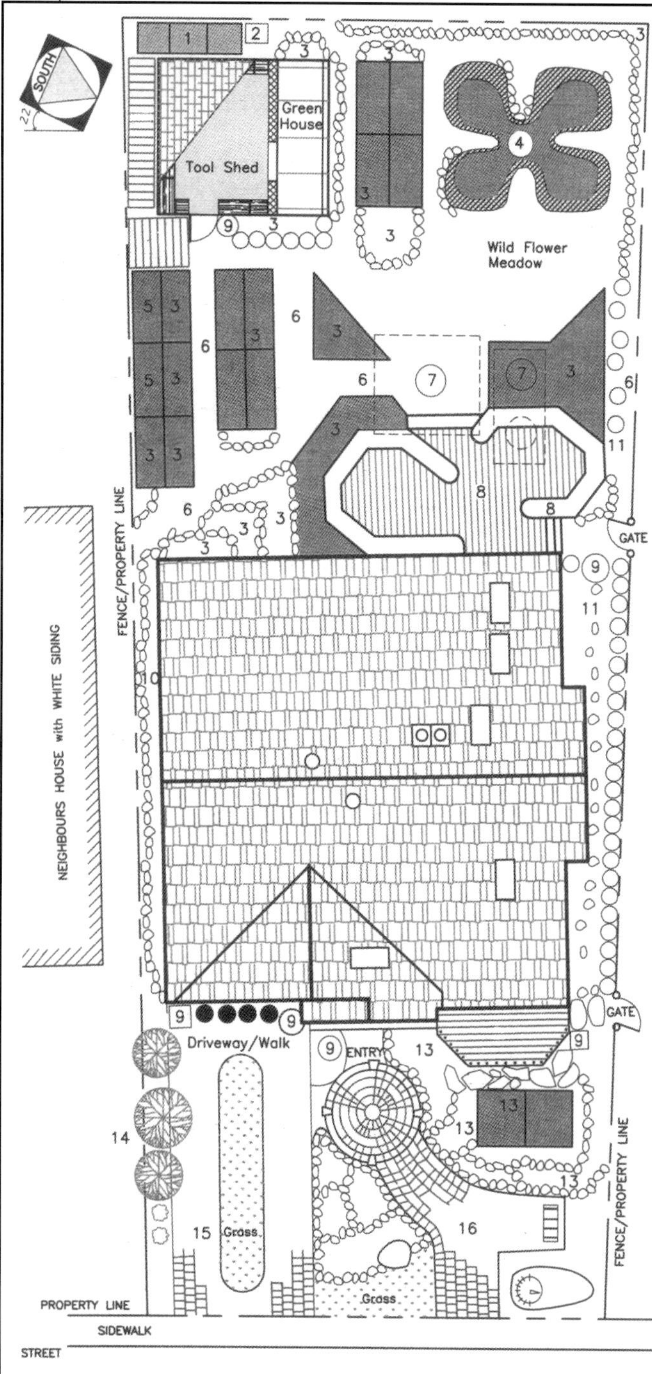

Throughout the lot, plants are selected to match the specific shade, soil and moisture conditions, like the shady northern raised beds (10), the sunny south garden (11), the local native garden (12), the west garden raised beds (13), and windbreak (14). The whole property is an organic permaculture garden in which plants are more than just attractive—they are edible or medicinal. The rich variety of species includes strawberries, raspberries, currants, Saskatoon berries, rhubarb, tarragon, lavender, lovage, and comfrey. About 75 per cent of the owners' food needs are met by their garden, including produce prepared for winter supply. Their growing season is extended by use of the greenhouse and wide windows in the house. The driveway (15) and front walkway (16) are surfaced with concrete pavers that allow water percolation and the driveway includes a grass strip between the car wheels that allows percolation.

The Ostrowski's garden is an excellent example of how working with natural processes can result in a beautiful, functional garden that gives back food and many other rewards.

Photo and plan: Jorg Ostrowski

Table 4-1: Comparison of some stormwater infiltration or recovery options

Stormwater Method	Advantages	Drawbacks
permeable hard surface (see Chapter 5)	• reduces runoff • allows infiltration	• some types, for example, loose aggregate, require cleaning, such as leaf removal, and weeding • the materials can be displaced during snow removal • costs vary depending on the material (refer to Chapter 5)
rain garden	• reduces runoff • allows infiltration • improves water quality • looks like planted beds or dry creeks • can beautify your property	• may give off odours and other problems if water is allowed to stand more than two days. Ensure proper rain garden size and soil characteristics. • may not be practical if space is limited, for example on small, urban lots or unsuitable soil and other site conditions • maintain vegetation
soakaway pit	• reduces runoff • allows infiltration • used for receiving water from roof downspout • does not take up space above ground	• system can clog. Use roof downspout filter that is inspected and cleaned in the autumn after the leaves fall to remove leaves and other debris before reaching the pit. If it overflows, unclogging is required. • maintain the grass and other vegetation over top • may not be practical if soil is unsuitable and if there are other unsuitable site conditions
planted swale	• directs water but allows some infiltration and runoff reduction and slowing	• potential for standing water. Ensure proper site characteristics and slope. Inspect it for water that is standing for over two days. • maintain the grass or other vegetation
rain barrel under downspout	• reduces runoff • recovery of water for watering your garden • low cost	• can become odorous. Use mesh screen or cover to keep out insects • can overflow. Use overflow pipe or hose at top. Direct overflow to planted areas, for example, a rain garden. • inspect and clean once a year

For winter, drain the barrel and outlet hose or pipe completely. Clean the barrel out once a year.

When Necessary, Convey Water

As discussed earlier in the chapter, stormwater should be directed away from your house and main outdoor activity areas, like driveways and patios. You can direct stormwater to desirable areas via grading, as described earlier, and/or using the following conveyance methods. Of these, planted swales are preferable in terms of reducing runoff.

a) Planted swales

A planted swale is a shallow channel typically covered with grass, although it can be planted with groundcovers that will tolerate moist to dry conditions. A swale can also be covered with permeable, loose, hard materials like river stone, pebbles or crushed rock, which can be used in

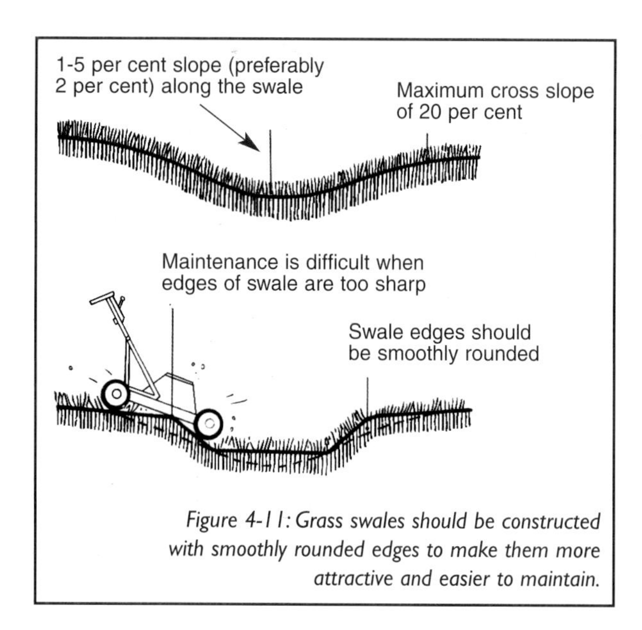

Figure 4-11: Grass swales should be constructed with smoothly rounded edges to make them more attractive and easier to maintain.

combination with plants. It is wider than it is deep and in many cases, is barely noticeable. Its function is to direct the surface water toward other parts of the drainage system, for example a rain garden or toward a front yard area that drains to the street. The swale also offers the advantage of absorbing part of the runoff into the soil, which tends to both reduce the volume and speed of runoff. When the slopes of the swale are gradual and rounded, it fits better into the landscape and is easier to maintain.

The slope along a swale should be 1 to 5 per cent preferably at least 2 per cent. Slopes that are too shallow may result in standing water while slopes too steep increase runoff speed and decrease infiltration. Ensure that conditions will support healthy, dense plant growth and if it is covered with grass, mow it so the height is at least 7.5 cm (3 in.).

If you don't have enough space for proper drainage clearance from the house (as described earlier in this chapter), put an impermeable lining under the swale to prevent water from soaking into the soil too close to the house. For tight spots like a side yard, talk to your neighbour and municipality about a shared swale on the property line.

If draining to or from an off-site location, check with your municipality for requirements.

b) Ditches

Ditches are deeper drainage channels than swales. Often used as a drainage solution for roads in rural areas, they are capable of carrying a greater volume of water at high speed than swales are. The potential volume of water carried is relative to the ditch's slope, dimensions and surface materials.

Rather than simple channels covered with grass, they can be planted with other species like perennials that tolerate wet to dry soils making the neighbourhood more attractive. Where possible, give them a meandering shape to resemble a natural stream. Check your municipality for regulations and approvals. A combination of plants and stone can help reduce erosion. Stones can be placed so that they create riffles that let low-water flows pass under the rocks, but absorb the erosive energy of high flows. This approach is even more impressive when it is shared by several neighbours, creating a planted network that links several lots together.

c) Catch basins and drain inlets

Catch basins and drain inlets are the entry points to an underground storm sewer. When they are necessary, consult a professional for advice. In the planning stage, well before connecting to an underground storm sewer network, obtain a permit from your municipality.

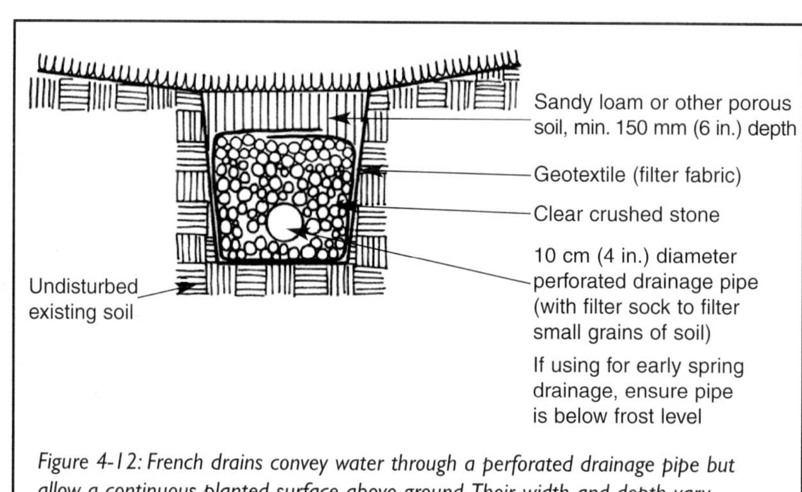

Figure 4-12: French drains convey water through a perforated drainage pipe but allow a continuous planted surface above ground. Their width and depth vary according to frost and soil conditions.

d) French drains

French drains convey water through a perforated underground drainage pipe but allow a continuous planted surface, like lawn, above ground. They consist of a perforated drainage pipe in a trench filled with clear crushed stone or free-draining granular material surrounded by a geotextile (filter cloth). The trench's width and depth below ground varies according to the soil and frost conditions. Place a layer of sandy loam or other coarse soil over this for the growth of grass or other plants above. Ensure that the drainage pipe has an adequate slope in the right direction. They can connect to a soakaway pit or other drainage element. If you are connecting them to the storm sewer network, consult your municipality for approval. (Figure 4-12)

Fig. 4-13: soaker hoses have pores that deliver water slowly and directly to the soil, which helps conserve water.

Water consumption doubles in summer due to outdoor uses like lawn and garden watering. It is estimated that as much as 50 per cent of outdoor use is unnecessary, like over-watering lawns or washing debris from driveways. Municipalities announce lawn-watering restrictions during very dry periods in summer due to difficulties keeping pace with the heavy demand. Water supply and treatment facilities built to meet this peak summer demand, have far more capacity than is needed for the rest of the year. Reducing your water use during the summer helps to avoid costly expansion of water facilities, diminishing the burden on municipal tax dollars.

IRRIGATION SYSTEMS

If you select plants that are adapted to the water available to them, very little irrigation would be necessary. Rain could meet the majority of the plants' water needs. So consider the range of plant options available before selecting an irrigation system, for example, xeriscapes. Chapter 6 describes this water-saving planting option and Chapter 7 provides several water saving tips for lawn, garden and trees.

Much of the water used for lawn and garden watering is wasted in that it is not taken up by the plant, but is lost to runoff or evaporation. Over-watering not only weakens the plant's root system, making it vulnerable to pests and disease, it carries nutrients down into the soil and away from the plant's roots.

The best approach to conserving water is to select plants that are suited to your local rainfall and property conditions, and to adopt the water-efficient design and maintenance techniques described in Chapters 6 and 7. If you do need to irrigate, the most water-efficient systems are those that apply water directly to the soil, like drip irrigation, since it's the roots that need the water. These are used for vegetable gardens, planting beds with perennials and shrubs, and under hedges and trees. For lawns, in-ground sprinklers are the most efficient option.

1. Above-Ground Sprinklers

Portable above-ground sprinklers are cheaper to buy and easier to install than the other systems but are not as water-efficient. Water can be lost to evaporation even before reaching the soil, or it can be wasted as runoff from pavements, depending on how you set up the sprinkler. Above-ground sprinklers also require supervision and time to move them around to reach the areas that need water. Select sprinkler types that best suit the size and shape of your planted areas, and place them carefully so they don't water areas that don't require irrigation, like pavements.

Watering with a hose and sprayer is also a common method of irrigation due to the low cost of buying and installing such a system. When they are held and pointed directly at the area that needs watering, they can offer more control than above-ground sprinklers, but they require more of your time. Again, much of the water is lost due to evaporation before it reaches the soil, for example when it is sprayed on leaves.

2. Soaker Hoses

Soaker hoses use considerably less water than an ordinary sprinkler or hose. (Figure 4-13) Place them at the base of plants in perennial and shrubs beds, vegetable gardens, hedges and under trees on or just below the soil surface. They are equipped with hundreds of tiny pores that emit water slowly. Evaporation is reduced since water is released directly to the soil and then it reaches the roots. These hoses also save time watering plants or planting beds since they don't need to be moved around. Another environmental advantage is that they can be made of recycled rubber.

Fig. 4-14b

Irrigation techniques:
a: timers allow you to automatically set the required watering time and amount, and limit it to the cooler times of the day in order to avoid evaporation and help to prevent over-watering.
b: portable above-ground sprinklers
c: in-ground sprinklers are formed of a network of pipes running through the ground with pop-up heads.
Photos: JMK Imag-ination

3. In-Ground Sprinklers

In-ground or pop-up sprinklers are fixed in place and supplied by a network of underground pipes. Nozzles designed to resist wind and reduce evaporation can help diminish water loss. Sprinkler heads can be accurately located to send water where it is needed and minimize water wastage. These systems can be very time and water efficient when they are properly planned. They can supply accurately the water each area requires and none of your time is needed to move the sprinklers around to reach each area. Initial costs are higher than for the previous options.

4. Timers and Sensors

The most time- and water-efficient systems are those connected to a timer or moisture sensor. The timer allows you to set automatically the required watering time and limit it to the cooler times of the day in order to avoid evaporation. The sensor or tensiometer evaluates the soil's moisture and the extent of the last precipitation. Therefore you avoid watering an area already moistened by a recent rain.

5. Drip Irrigation

Drip irrigation systems reduce evaporation water loss to a minimum by delivering water directly to the soil. These systems send small amounts of water very slowly to the roots of each plant through emitters fitted on flexible plastic tubing. Drip irrigation saves significant amounts of water and time compared to other systems, although the capital cost can be higher.

WATER PONDS

The water pond was once a luxury that was not accessible to most people. Today, materials and construction techniques have been refined to allow the integration of water ponds into almost any situation (Figure 4-15). Since then, we are witnessing a proliferation of ponds and fountains. Some of these ponds are strictly aesthetic while others harmonize with the natural environment and are rich with flora and fauna.

Design considerations for ponds:

1. Plan your pond like a diverse, natural ecosystem

A water pond can be a mini-ecosystem in which each living organism has a place. Plants create food, shelter and oxygen for animals like frogs and fish that in turn give back nutrients to the plants. Moist, aquatic environments support a wide range of fauna and flora and are important for the survival of other ecosystems. If you want to attract birds and other wildlife to your garden, ponds are an ideal feature. When selecting species, remember that ponds with a more diverse range of plants and animals tend to have fewer problems, like mosquitoes. Some species, such as frogs and toads, will naturally colonize your pond, depending on the proximity of natural areas to your home.

It is easier to achieve a clear pond when stormwater runoff from your yard does not drain into it. Runoff can introduce excess nutrients that can encourage algae growth as well as soil particles and debris. Certain pond lining materials are impermeable and designed for closed-loop systems, where the water does not infiltrate into the soil and get replenished by stormwater runoff. However, with materials like clay, you can create a more open system that collects runoff by locating your pond at a point where stormwater collects naturally. This type of pond fluctuates with the rhythms of natural cycles. If you do have an open system that accepts stormwater runoff, ensure that you have enough

Fig. 4-15: This small urban front yard incorporates a water pond.

space to enable it to be at a safe distance from the house to avoid foundation drainage problems. Refer to the section on clay ponds later in this chapter. It will also be important not to apply pesticides or concentrated fertilizers (synthetic or non-synthetic) on adjacent planted surfaces, like lawns, that will be carried to the pond when it rains.

2. Locate your pond in partial shade

Choose a location that receives medium shade, or about four to six hours of sunlight per day. To minimize shade and fallen leaves in your pond, try to avoid locations under trees. You should also choose a location with access to a water supply for initial filling of the pond and topping up due to evaporation losses, and to a power supply if you have an electric pump or other equipment. Once you find the best location, use a garden hose or pegs and string to mark-out the shape on the ground. Check for underground utilities before you dig.

3. Determine correct depth

The generally recommended minimum depth for a small pond is 50 cm to 1 m (20 in. to 3 ft. 3 in.). That way the plants will be able to survive winter in regions that experience freezing. Some municipalities require fencing around ponds of certain depths, so always check municipal bylaws at the beginning of the planning process.

When a pond is not deep enough, in hot weather, water temperatures can climb rapidly, which encourages the growth of algae. The growth of excess algae can increase oxygen demand, which can kill fish and other aquatic organisms. Installing the pond in partial shade where it will only receive morning or evening sun will help to avoid the problem. By providing plants with floating foliage, such as water lilies, the leaves will help keep the water cool and fresh. They should cover at least 50 per cent of the surface of the water.

4. Don't forget about aeration

Oxygen is needed to maintain water quality and life forms in the pond. The use of oxygen-generating plants and filtering plants improves the water quality in the pond by oxygenating the water and reducing superfluous nutrients, greatly reducing algae growth. Also, mosquitoes are only attracted to standing water, so keeping water circulating and aerating is needed to reduce this problem.

If plants and other life forms do not provide sufficient aeration, water will have to be aerated mechanically. Water circulating and aerating pumps, fountains and waterfalls are installed in ponds, usually at the deepest point, to perform this task. Pumps are used with waterfalls and fountains that both help to aerate the water and create the soothing sounds of falling water. (Figure 4-16)

Fish, frogs and toads are an integral part of a pond's ecosystem, and are helpful in feeding on mosquitoes. Some fish, like goldfish can proliferate quickly and become a problem, while others, like Japanese carp, can quickly devour your aquatic plants. It is preferable to introduce fauna that naturally occur in your area, rather than exotic ones. Talk to specialists for advice on the type and quantity of fish to suit your pond size and conditions.

Fig. 4-16: Waterfalls add beauty, movement and soothing sounds to your yard, in addition to improving water quality and aeration in the pond.

To help reduce mosquito breeding areas in your yard, eliminate stagnant water, for example by emptying and cleaning birdbaths twice weekly until fall frost. Mosquito larvae can develop in water that has been standing for more than four days. Ensure that you have adequate water aeration and circulation in your water pond.

5. Choose the right type of structure and material

a) Prefabricated Ponds

Prefabricated ponds are usually made of rigid plastic or fibreglass and are available in various shapes and sizes. They are simpler to install but less flexible in terms of size and shape than the other options described below. These ponds are not used to recover stormwater runoff from the rest of the yard. They are used to create small, decorative closed-circuit ponds. Installation is similar to that of impermeable pond liners shown in Figure 4-17.

b) Impermeable Pond Liner

After prefabricated ponds, impermeable pond liners are the easiest to install. They hold the water like a swimming pool liner. Since they are flexible, they can be installed and adjusted to give any shape. They also tend to be cheaper than prefabricated ponds. The impermeable membrane is installed on a bed of sand or a felt underlayment. The liner should be covered with smooth rocks, pea gravel and pebbles to provide UV protection and other benefits, which are described below. This type of pond is a closed circuit system that is not used to collect runoff from the rest of the yard (Figure 4-17).

Water may accumulate under the liner causing it to buckle and lift. If you are installing the liner on clay soil or you notice water pooling up while you are digging, install a pressure relief valve in the pond bottom through the pond liner to let water that may build up under the liner enter into the pond, rather than accumulating underneath. The valve prevents water inside the pond from escaping to the outside. Depending on the size, more than one valve may be needed.

c) Clay

For very large properties, like in rural areas, clay may be used as a sealant (Figure 4-18). This type of pond can receive a supply of water from stormwater runoff. Look for areas on the property that are already wet or that collect water

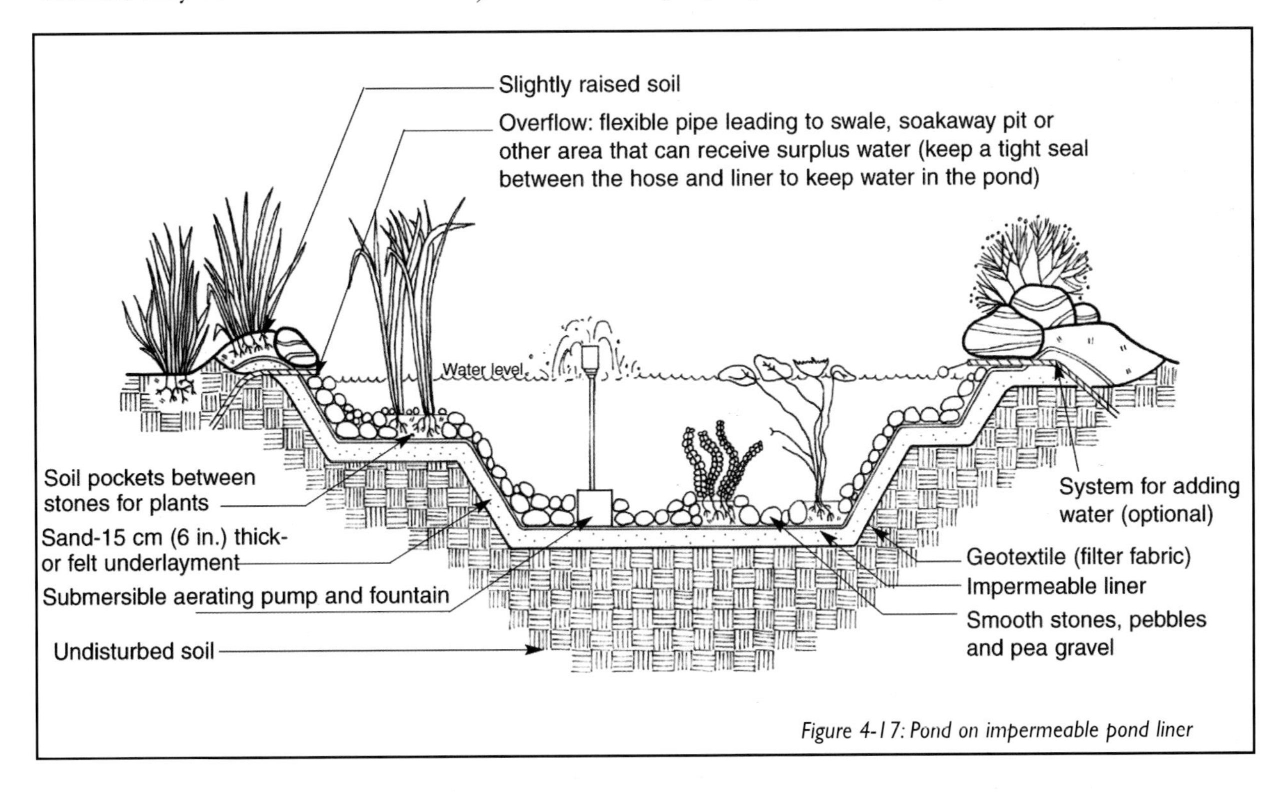

Figure 4-17: Pond on impermeable pond liner

Figure 4-18: Pond on layer of clay

naturally. Also ensure it is located far from the house to avoid foundation drainage problems.

Clay is heavier and harder to work with than other materials like impermeable liners and prefabricated plastic ponds. It does not suit all soils. If placed on sandy soil, for example, the pond may empty from the bottom too rapidly and dry out. Clay keeps plants in place, holds nutrients for plant health and provides habitat for aquatic life forms. Another advantage is that it is not a manufactured product and is often locally available, perhaps even on your property, which helps reduce costs and transportation energy consumption. With this natural, flexible material, you can create a more complex and dynamic pond contour for fauna and flora. It will be easy to recreate a bank and even a wet area where frogs and dragonflies will settle. Since water levels will fluctuate, plants at the edge should be adaptable to dry and wet soil conditions. If you already have a naturally created wetland, it is best to leave it alone. Check with your local municipality or environmental authority before working in or around a wetland.

6. Rocks, pebbles and soil on the pond bottom

Placing smooth rocks, pebbles and soil on the pond bottom serves several functions. It provides soil medium for the plants, holding them in place, and shelter for fish and amphibians. Bacteria also colonize these surfaces and help break down organic matter. Smooth rocks, pebbles and soil also hold down pond linings, provide less slippery surfaces for walking and protect linings against UV deterioration. They also give the pond a more natural appearance.

7. Waterfalls

In nature, water movement and aeration are important because they bring oxygen to the aquatic environment and help reduce mosquitoes and algae problems. Drops, waterfalls and currents help water to mix with the air, oxygenate and limit stagnation. Cool oxygenated water limits algae propagation and mosquitoes. Where possible, install the waterfall between two naturally occurring levels. To contain splashes and limit water loss, the width of the receiving pond should be approximately twice the height of the fall.

8. Choose the right plants

Plants oxygenate the water and improve water quality. They also block sun, thereby reducing algae growth. Select plants that suit the varying conditions found at different locations in the pond. For example, locate plants with floating leaves, like water lily and water hyacinth, in the deeper parts of the pond away from the splashing water from waterfalls or fountains. Locate emerging leaf plants, like cardinal flower and cotton grass, in shallower water near the pond edge. Refer to the Plant List for Water Gardens (Table 4-2) and talk to a specialist for

Table 4-2: Plant List for Water Gardens

Plant Group	Roles	Name	Height - cm (in.)	Foliage	Flowers
sub-merged	oxygen -ation,	*Elodea Canadensis* / Waterweed	100 (39)	• greenish beige cylindrical stem	• minuscule white flowers at the end of a long very thin stalk
	filtration	*Myriophyllum sibiricum* / Common Water Milfoil	120 (47)	• whorled and very finely cut leaves	• very small stalkless flowers
		Ceratophyllum demersum / Coontail		• dentate, stiff and pungent leaves	• flowers in leaf axils
floating foliage	oxy-genation,	*Nymphaea* / Water Lily	45 to 150 (18 to 59)	• floating leaves	• varied colours
	shade	*Nuphar* / Pond Lily	A few millimetres	• floating leaves	• yellow flowers, July to August
floating	shade, filtration	*Lemma minor* / Duckweed	A few millimetres	• small oval leaves	• minuscule flowers, hard to see
emerging foliage	shade	*Acorus calamus* / Sweet Flag	60 to 75 (24 to 30)	• dark green	• of no interest
	filtration	*Alisma plantago-aquatica* / Water Plantain	30 to 45 (12 to 18)	• oval leaves	• delicate pink and white flowers. Flower stalk may reach 90 cm (35½ in.)
		Caltha palustris / Marsh Marigold	15 (6)		• white spathe
		Eriophorum / Cotton Grass	30 (12)	• narrow linear leaves	• napped flower from May to June
		Iris versicolor / Blue Flag	55 (22)	• linear leaves	• veined blue violet flowers with yellow and white throat
		Juncus effuses / Common Rush	250 (98)	• forms clumps of green stems	• small, inconspicuous
		Menyanthes trifoliata / Bog Bean	30 (12)	• thick stems, leaves in groups of three	• white flowers, May-June
		Orontium aquaticum / Golden Club	15 to 25 (6 to 10)	• thick extended leaves	• yellow and white floral stalk
		Pontederia cordata / Pickerel-Weed	45 to 75 (18 to 30)	• lance shape	
		Sagittaria latifolia / Broad-Leafed Arrowleaf	45 (18)	• sagittate leaves	• mauve flowers, June to October white flowers in summer
		Scirpus acutus / Hard-Stem Bulrush	Up to 300 (118)	• few and short, found at the base	• terminal purplish brown spikelets
		Scirpus lacustris / Great Bulrush	90 to 120 (35 to 47)	• round stems	• brown flowers in herring-bone pattern, June to July
		Typha latifolia / Common Cattail	100 to 270 (39 to 106)	• wide leaves	• dark brown, August to September
		Lobelia cardinalis / Cardinal-Flower	90 to 120 cm (35 to 47)		• spectacular red flowers
		Typha angustifolia / Narrow-Leaved Cattail	120 to 150 (47 to 59)	• narrow leaves	• dark brown herring-bone shape, August to September

Adapted from: *Fleurs, plantes et jardins, les cascades et les jardins d'eau*, volume 2, number 2, May 1991

Life in the Pond: Pollard/Meredith residence, Cantley, Quebec

After installing a pond at his residence in Cantley, Quebec, Doug Pollard could soon hear the cheerful choruses of frogs attracted to it. The pond took on a life of its own, as toads, frogs and plants, like cattails, colonized it from surrounding habitats. What Doug and Charlene Meredith like best about their pond is seeing and hearing the changes in different seasons and from different angles. The waterfall also adds movement and life. These dynamic elements make the pond a pleasure to the eyes and ears.

Before

Photos: Doug Pollard

After

The pond bottom is an impermeable pond liner covered with locally available soil and rocks along the edges. Doug took advantage of a naturally occurring grade change for a waterfall. Water is pumped using a submersible pump at the deepest point, about 1.2 m (4 ft.) deep, in the lower pond up to the upper pond where it falls over a 1.2 m (4 ft.) drop over carefully placed rocks where it lingers in a shallow pool, then drops another foot back into the lower pond. Doug relies on the waterfall and natural processes in the pond for water quality and has no supplemental filtration system. He planted water hyacinths and various species of iris in the water, as well as dogwoods and assorted perennials along the slope, outside of the water. Maintenance is limited to scooping leaves off the surface in the fall and bringing submerged pots of water lily and bamboo into the house for the winter.

more details. Plants can be installed directly into the soil or rocks on the pond bottom or placed in pots. After planting, cover the soil with smooth rocks or pebbles.

9. Filtration and maintenance

Aquatic plants and other life forms will help to filter the water, making it clearer and healthier. Removing fallen leaves in the autumn will also help. For a list of filtering plants, consult the Plant List for Water Gardens (Table 4-2).

If there are not enough filtering plants, a filter may prove necessary. Mechanical filtration strains out solid particles and algae and may need to be cleaned regularly. Biological filtration uses beneficial bacteria to break down waste. Some systems combine both mechanical and biological filtration. You can set your own criteria for the degree of water clarity you prefer. It may be time to review these criteria with a more environmental eye, because clear water is not necessary to maintain the equilibrium of the pond ecosystem.

10. Prepare for winter

In autumn, remove fallen leaves from the pond. Just before leaves fall, you can place netting over the pond and remove it afterwards to make this task easier. If you have selected only hardy species that can overwinter outdoors, you will save time having to bring pots of non-hardy plants indoors for the winter. Cut back the above-water parts of hardy water lilies and bring pots of non-hardy ones indoors for the winter. Remove the pump from the pond for the winter and store it in a bucket of water indoors. For tips on over-wintering fish, talk to a pond expert.

LIVING BY WATER

Healthy shorelines are a vital edge where plants and wildlife, both in the water and on the land, find the resources they need. They can act as wildlife corridors, linking larger habitat areas. We too depend on healthy shorelines. If your home is adjacent to a creek, river, lake, wetland or other waterbody, the water quality and habitat is affected by how you landscape. The natural mix of trees, shrubs and grasses along a shoreline help to stabilize slopes, slow down and reduce runoff, diminish erosion and soil loss and filter out soil particles and other substances from the water, acting like a buffer and protecting habitat. Dramatically altering their natural state of shorelines, for example by removing vegetation, has multiple negative impacts.

Here are a few tips you can use to help preserve healthy habitat along a buffer near the water's edge:

- If you have the space, it's best if this buffer is at least 30 m (12 ft.) wide, measured back from the high water line. Check the regulations.
- If you already have trees, shrubs, groundcovers and aquatic vegetation, particularly those that are native to your area, leave them in place.
- Plant new ones if there aren't any. Look for native species that naturally grow at the water's edge in your area when developing your plant list. This provides shady habitat in and beside the water. Refer to Chapter 3 for tips on planting on slopes, for example soil bioengineering. These planting techniques are used to stabilize slopes above the low-water level (areas that are only seasonally flooded or above water). Erodable areas that are permanently underwater should be protected with rock. Also refer to Chapters 6 and 7 for planting tips.
- A healthy shoreline mimics the complexity of nature in terms of the different plant species, ages and sizes from low groundcovers to tall trees. Rotting material like leaves and branches provides nutrients for plants and habitat for wildlife on land and in the water. Tall plants shade and protect smaller ones and as they die, make room for new ones to grow.
- Eliminate your use of fertilizer and pesticides near the water.
- Leave rocks and natural debris like logs and fallen trees in place. These provide protected nooks that are essential to many species. You can even put these in place where they don't exist.
- Cover bare patches of soil that can lead to soil erosion and the spread of invasive plants that can crowd out native species. Also try to use only non-invasive plants on your property and avoid dumping yard waste containing invasive plants near natural areas.
- Various agencies have regulations and require approvals for work beside a body of water, including your local municipality, the federal Department of Fisheries and Oceans, and provincial and territorial ministry responsible for the environment or natural resources. Check these requirements during the planning phase of your project.

For more information on healthy shorelines, refer to the list of resources at the back of this book.

Shoreline revegetation: Lowens residence, Gananoque, Ontario

Ed Lowens re-established vegetation on this Lake Ontario shoreline at his summer home in Gananoque, Ontario. The sandy beach and surrounding area had been cleared of vegetation. The cattails, water lilies, willows, ash, pine and spruce that he planted have established and multiplied to form a lush, green and healthy water front. Now it is a home to red wing blackbirds, a great blue heron, muskrats and many families of ducks and Canada geese.
Photo: Ed Lowens

Chapter 5

HARD SURFACES

Hard surfaces, such as driveways, patios and walkways, can occupy a large portion of your property. They are used because they are solid, durable and slip-resistant. These characteristics let them stand up to repeated and intensive usage.

Surfaces covered in paving are often impermeable, which results in three significant consequences for the environment.

1. First, because they prevent water from slowly soaking into the water table, they cause rapid runoff of surface water toward the storm sewer system and toward natural water bodies. This means that pollutants, such as road salt, oils and other substances can affect water quality and aquatic life. High volumes of runoff also place stress on municipal water systems.
2. Secondly, impermeable pavements prevent water from reaching plant roots, therefore limiting the benefits of having healthy trees and other vegetation, which provide shading and beauty on your property and in your community.
3. Finally, pavements make urban areas feel hotter in the summer, since they absorb solar rays and re-radiate them as heat. Plants, on the other hand, transpire water vapour through their leaves, reducing heat as the water rises into the air. Runoff from pavements removes water and reduces evaporative cooling in summer heat.

Carefully consider what you need on a regular basis and limit paved areas. This leaves more

Driveway Conversion: Sanders-Fisher residence, Ottawa

To reduce runoff and beautify the property, the homeowners reduced the size of their driveway by over 60 per cent after assessing their parking needs. The asphalt driveway shown in the "before" photo was replaced with hardy low-maintenance plantings and precast concrete pavers. The asphalt was removed and sent to a recycling facility, which saved the homeowners the tipping fees and prevented the materials from going to the landfill. The aggregate base underneath the old asphalt was reused under the new concrete pavers. It was initially collected and stockpiled on-site, while the area under the new pavers was excavated to the required depth for the deeper drainage base under the concrete pavers. Aggregate material from the areas that were to be planted was replaced with soil that included compost.

Before

After

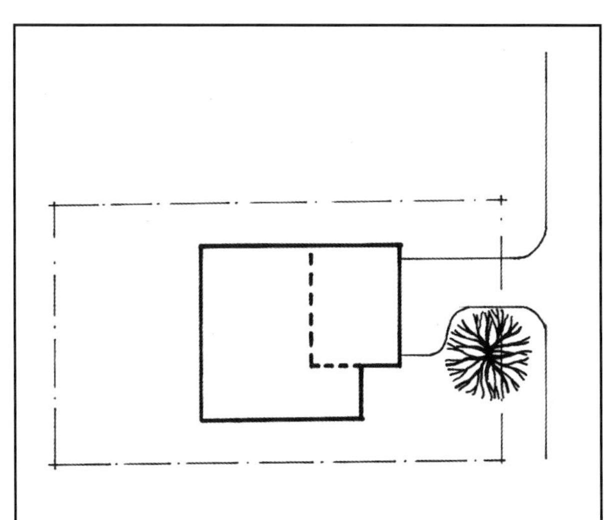

Figure 5-1: If you have a two-car garage, you can minimize the surface area of pavement by designing a single-entry driveway that widens toward the garage door. In the example shown, the layout enabled the tree preservation.

room for plants! Where you need hard surfaces, favour permeable materials that let water soak into the ground.

DRIVEWAYS

When planning your driveway, try to:

1. Reduce paved surfaces to the acceptable minimum required to meet your needs
You can do this by carefully assessing your daily needs. The same hard surfaces can be used for different activities. For example, a driveway can be used for basketball, hopscotch or other games. Take advantage of driveways and patios for foot traffic.

To minimize driveway space, ask yourself about your real parking needs. In Chapter 2, we talked about knowing your needs. If you are using your driveway to park cars, think about how many will be parked on a day-to-day basis. You can reduce the width of the parking space by widening the area only where you park a second car or by parking them one behind the other (Figure 5-1). Is street parking available for visitors?

If your driveway is mainly used to access your garage, it doesn't need to be more than 2.4 m (8 ft.) wide for the average car. Keep the length to a minimum by making it as direct as possible. The size of your parking spot depends on several factors including the size of your car, obstacles adjacent to the spot, such as fences or walls, and how wide a space you need to get in and out of the car. A good rule of thumb for a parking space is about 2.6 m (8½ ft.) wide by 5.5 m (18 ft.) long for an average sized car. For wheelchair access the parking space width should be at least 3.9 m (12 ft. 9 in.) wide. Check and adhere to your municipality's requirements on the size, location, materials, grading and visibility of driveways and parking spots on your lot.

2. Choose permeable surfaces to let stormwater soak into the ground
One approach to reducing the amount of paving is to install two strips of paving for vehicular movement, and planting grass or a groundcover that tolerates dry conditions for the space between the strips. The strips should be at least 60 cm (2 ft.) wide (Figure 5-2).

Figure 5-2: Rather than having an extensive swath of asphalt, this home features a driveway made primarily of grass, with the exception of narrow bands of precast concrete pavers. This helps stormwater soak into the soil and reduces the amount of runoff being sent to the storm sewer. Low-growing ground cover plants could have also been used to save on mowing.

In addition to limiting the size of driveways, permeable materials can be used, such as precast concrete pavers that are designed to be permeable or loose aggregate materials, such as crushed brick. Refer to Table 5-1 to compare the permeability of different materials.

3. For intensive needs use durable materials that are easy to use and maintain

Planted or grassy areas cannot withstand heavy and repeated vehicular traffic, so hard surfaces are needed. Some materials are more durable and resistant to intensive usage than others. Refer to Table 5-1 for a comparison chart of materials, including durability and maintenance needs, and to the following sections for technical and installation details.

Some materials also require more maintenance than others. Uneven or loose surfaces, like natural stone, decorative pebbles and mulch require special attention when removing the snow. It becomes a delicate operation trying to avoid destabilizing loose materials or pushing some off the site. In the case of decorative pebbles, for instance, in the spring you'll need to rake the stones pushed onto plant beds and the lawn and put them back in place. Weeding will also be needed for these types of surfaces.

Designing Accessible Routes

For wheelchair access where there is a change of elevation, for example, from the driveway to the main doorway to the house, the path should have no more than a 1:20 (5 per cent) slope and have a cross slope not steeper than 1:50 (2 per cent). The surface material should be slip resistant, smooth and firm enough to accommodate a wheelchair. If there is a drop-off at the edge of the path, you should provide a 10 cm (4 in.) high curb edging. (Figure 5-3) The path should be at least 1.2 m (4 ft.) wide. If the path exceeds a 1:20 slope, it should be designed as a ramp, which generally can have a slope of up to 1:12. Details on ramps and other aspects of accessible routes are covered in CMHC's publication *Housing for Persons with Disabilities.*

Figure 5-3: This entry provides both stairway access and wheelchair access to the main doorway. The walkway is sloped for wheelchair access, with a 10 cm (4 in.) high border.

GROUND-LEVEL PATIOS

Patios can provide a comfortable place to eat, chat, read a book or just sit and observe nature. But as with driveways, try not to make these spaces larger than required to meet your needs.

Plan your patio according to your needs and adjust the location, shape and materials to the site and intended uses. Planning for a patio is similar to designing a deck. For more details on patio design, including space requirements and best locations, refer to the section on decks later in this chapter.

Refer to Table 5-1 for a comparison of costs and relative advantages of various patio materials, and

Figure 5-4: A patio for a small space

Table 5-1: Choice of hard surface material

Surface material	Durability	Initial cost*	Repair and maintenance	Environmental considerations**
loose aggregate (such as, decorative pebbles, pea gravel, crushed stone, crushed brick)	fair	low	- weeding required - regular cleaning of adjacent surfaces because these materials tend to shift. Edging can help. - snow removal difficult because these materials are loose - surface cleaning (such as fallen leaves) and levelling of small depressions after the winter	- permeable—allows water infiltration and reduces runoff - reusable - can be made from recycled materials (such as crushed concrete or brick)
wood chip mulch	low (for low-use foot-traffic)	low	- mulch to be added each year since it decomposes - regular cleaning of adjacent surfaces since it tends to shift - snow removal difficult because mulch is loose - weeding required - some mulch can be blown away in the wind	- very permeable—allows infiltration and reduces runoff - keeps tree trimmings, wood scraps out of landfill sites—recycled - old mulch can be composted - renewable - least durable and wear resistant compared to other materials - for low-use footpaths
natural stone	excellent	high	- replace damaged stones - snow removal difficult since the surface is rough - weeding may be needed between stones - very durable	- less runoff when gaps between stones are wide and filled with permeable material, like sand and plants - reusable - recyclable - very durable
reinforced grass rings	average	average	- regularly mow and maintain the grass - snow removal difficult since the surface is rough - grass compacts and must be replaced if overused	- permeable—improves water infiltration and reduces runoff - increases plant surface, cooling city heat in summer - recyclable - some products made of recycled plastic but new plastic manufacturing is energy-intensive
precast concrete pavers	excellent	average	- very durable if properly installed - localized repairs are possible because you can easily replace single damaged pavers - firm, even surface—good for mobility and intensive use	- reusable, recyclable - very durable material - somewhat to very porous, depending on the type (some pavers are designed to maximize infiltration) - energy-intensive manufacturing process
concrete	excellent	average	- very durable, if built properly - easy snow-removal - subject to cracking - firm, even surface—good for mobility and intensive use	- increases stormwater runoff and reduces infiltration, unless using porous concrete - recyclable - durable material - energy-intensive manufacturing process
asphalt	good	low to average	- repair cracks - durable if properly installed - easy snow removal - firm, even surface—good for mobility and intensive use	- increases stormwater runoff and reduces infiltration, unless using porous asphalt - energy-consuming manufacturing process - fumes during installation - recyclable
decks (more details in Table 5-3: Choice of Deck Materials)			- more in Table 5-3	- permeable if water can drain through gaps between the deck boards and if permeable surface underneath - reusable - wood is renewable - more in Table 5-3

*More durable materials can be less costly in the long run, since they have lower maintenance and replacement costs.
**Transportation energy is reduced by choosing materials harvested and manufactured locally. Environmental impacts of these materials depend on raw material harvesting and regeneration practices.

to the sections on installation and technical details later in this chapter. Common materials for patios include natural stone, precast concrete pavers, poured concrete and precast concrete slabs.

Decks are normally used where it is desirable or required to be raised above the ground, whereas patios are built at ground level. Views, slopes and soil depth are among factors to consider. For example, if you want to create an intimate space low to the ground that prevents neighbours from seeing you over a hedge or fence, consider a patio. If the bedrock is near or at the surface or the site is steeply sloped, or if you want to see over an obstacle, consider a deck.

WALKWAYS

Although driveways can take up a lot of space, the majority of movement on the property is nevertheless made on foot. Certain behaviour and traffic patterns on your property are predictable, making it possible to locate walkways effectively to withstand the degree of foot-traffic you anticipate. Once you have analyzed these patterns, choose the location, layout and materials for your pathways.

Figure 5-5: The material you select should suit your design. For example, the square precast paving slabs are better suited to the geometric design with very clean, trimmed forms in the upper photo.

Whereas, the irregular shapes of the natural stone in the lower example are better suited to the curving forms in this unmanicured backyard.

Follow these objectives:

1. Use durable materials for intensive needs and looser materials for less frequently used paths

The best materials for walkways are determined by the intensity of the anticipated usage. Walkways are often less intensively used than patios. So it is easier to choose more irregular or loose materials, like decorative pebbles, mulch or natural stone. Refer to Table 5-1 for a comparison of materials and to the following sections for technical details.

Main walkways, like the link from your home's main access to the street or the driveway, are usually covered with a durable material, like precast concrete pavers, since they are used more frequently. Secondary walkways, like the link from the front yard to the backyard may be used less frequently, so you can use more irregular surfaces like natural stone or stepping stones or loose materials like crushed brick. There may be other activities and destinations in your yard that are even used less frequently, like a path through a woodland or wildflower meadow garden. These can be covered with loose materials such as mulch or pebbles.

An effectively designed foot-traffic pattern prevents users from taking shortcuts. Lawns or plant beds worn by pedestrians suggest that you should re-examine the placement of your walkways. To correct the situation, modify the shape of the walkway to guide the walkers. Well-located plant beds that frame the walkways create interesting views and guide the users.

2. Reduce the width and number of walkways to the acceptable minimum

To limit the space required for pathways, try to take advantage of other hard surface areas, like decks and patios for foot traffic.

Paths for pedestrians should be wide enough to satisfy the number and type of users anticipated. Walkways are usually at least 75 cm (30 in.) wide. For ease of travel by wheelchair, the walkway should be at least 1.2 m wide (4 ft.) and the lengthwise slope should be up to 1:20 (5 per cent) unless the path is designed as a ramp (also refer to "Designing Accessible Routes" earlier in this chapter).

3. Choose permeable materials

Walkways can use the same materials as patios or driveways, but more permeable, loose materials, like crushed stone and mulch, are more appropriate for walkways because of lower intensity of use.

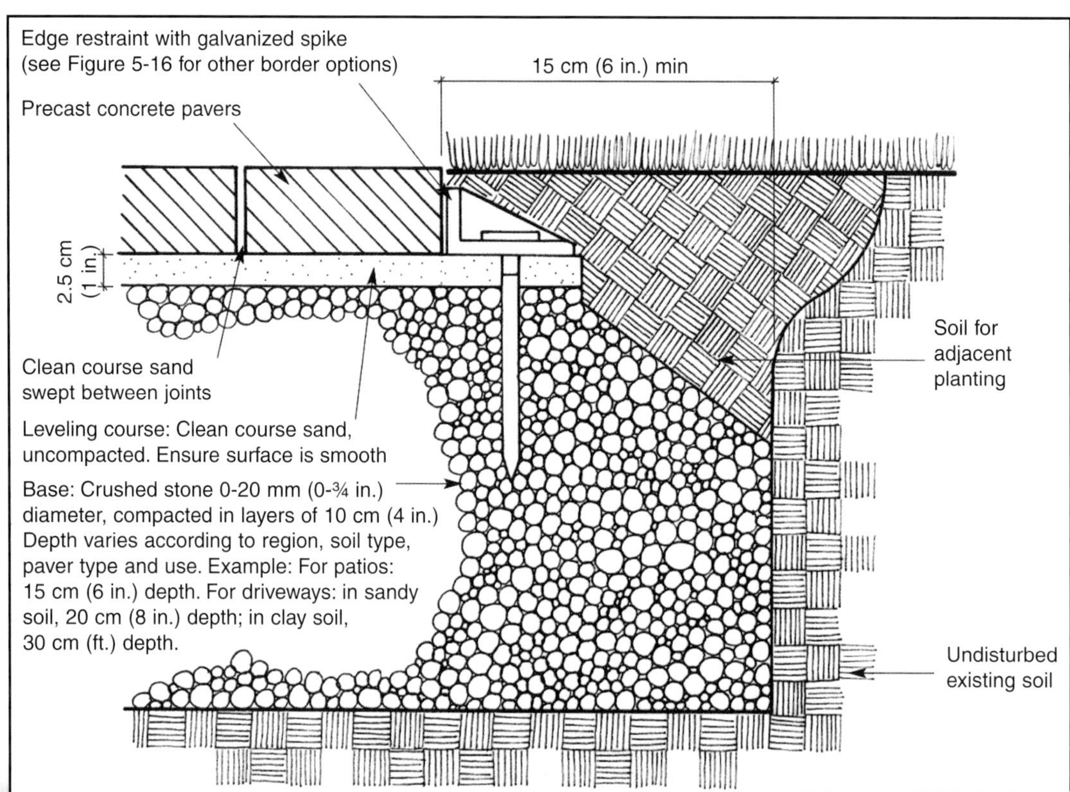

Figure 5-7: Precast concrete pavers

Figure 5-6: Precast concrete pavers used for a patio

Figure 5-8: Compact the base every 10 cm (4 in.) layer before placing the leveling coarse and pavers.

CHOICE OF HARD SURFACE MATERIALS

Precast concrete pavers

A vast array of colours, sizes, shapes and textures of precast concrete paver are available to fit your taste and the style of your home. Several companies offer pavers that imitate natural stone or old, weatherworn paving stones that give an aged look right from the outset. Different pavers can be combined to make attractive patterns.

The success depends on the base preparation. In regions where ground freezes, ensure that you have the correct depth for the base and that the base is compacted. A solid foundation that drains well avoids water accumulation. Water that freezes under the paving swells and deforms its surface. A poor base often leads to costly repairs. Figure 5-7 and 5-8 show examples of installation methods. Consult manufacturers brochures for paver types and installation details. To reduce tripping hazards, pavers should be set to minimize surface irregularities. In regions where ground freezes, precast concrete pavers have replaced clay brick as paving. Since they are porous, bricks absorb water when they are installed on the ground and burst when subjected to freezing, limiting their life span.

The gaps between precast concrete pavers allow some water to percolate into the soil, depending on the size of the spaces between the pavers and the type of paver. Some pavers are designed specifically to maximize stormwater infiltration, for example by enabling larger gaps and channelling of water into the gaps. These pavers are available under different brand names and allow a significant percentage of stormwater to infiltrate. Stormwater infiltration can be maximized by using a course clear crushed stone, 38-50 mm (1½-2 in.) in diametre, that is free of fine particles in the foundation.

Other environmental advantages include the fact that they are manufactured in plants where the concrete quality standards are carefully followed to increase their durability. Moreover, damage can be easily repaired by replacing individual pavers as needed, so there is less wastage in replacing the whole surface, helping to make repairs easier than for non-modular surfaces. Another advantage is that they can be removed and reused elsewhere. However, cement manufacturing is an energy-intensive process. Finally, precast concrete pavers also provide a firm surface that is good for mobility and intensive use.

Asphalt

Asphalt is frequently used for large surfaces with intensive use, such as driveways. It provides a firm, even surface that is good for mobility and intensive use. With a low installation cost, it requires little maintenance afterwards. Cracks are generally caused by inadequate base preparation or by deep frost action. The depth of the crushed stone base varies depending on the region and local standards, but one rule of thumb is 5-10 cm (2-4 in.) for sandy soil, 10-15 cm (4-6 in.) for loam and 15-20 cm (6-8 in.) for clay soils. It is always important to fully compact the crushed stone base, as well as the asphalt. Properly constructed asphalt surfaces often last 10 years before they have to be resurfaced.

Asphalt processing is energy-intensive. However, asphalt can be recycled if you decide to remove it. Look in your Yellow Pages™ for companies that recycle asphalt. This will save you money on tipping fees and keep the asphalt out of the landfill.

Another environmental drawback is that regular asphalt is impermeable and does not allow water to soak into the soil. However, ask about the availability of porous asphalt. It uses larger sized stone particles than regular asphalt, and lets water drain through the voids in the paving, helping to reduce runoff and increase infiltration.

Poured concrete

Concrete is adaptable to just about any shape before it hardens and provides a firm, surface that is good for mobility and intensive use. It is usually more expensive than many of the other options. It is, however, a very durable material that requires only a minimum of maintenance over the long term if installed properly. It can crack during freeze/thaw cycles if installed incorrectly.

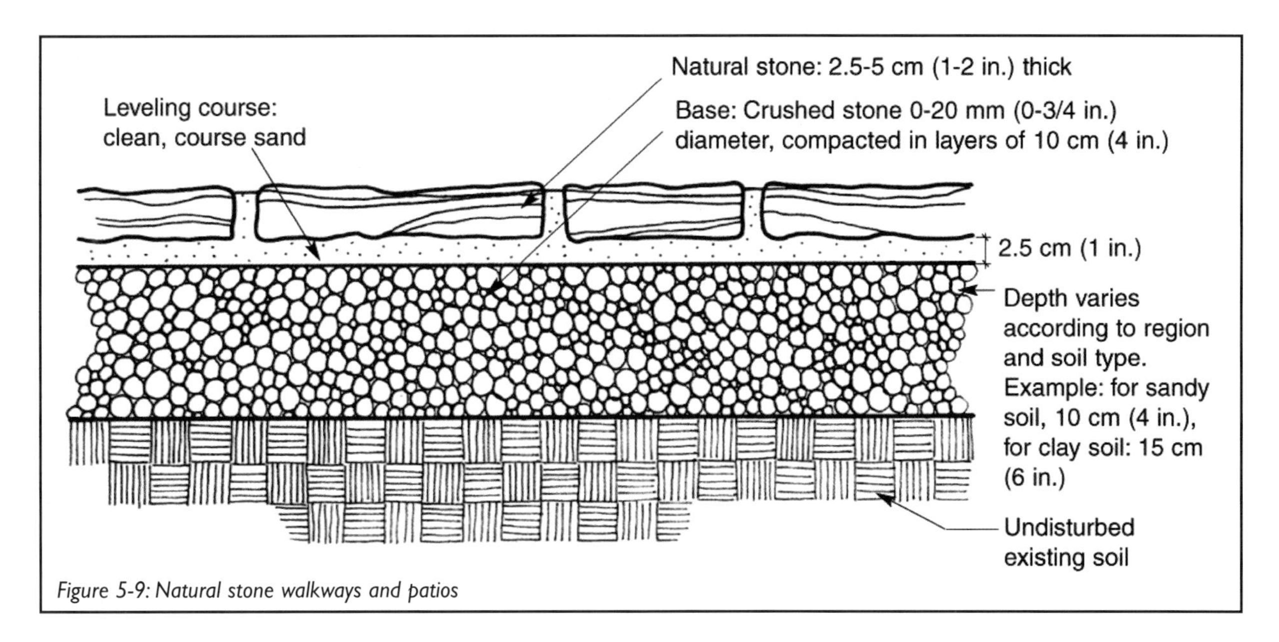

Figure 5-9: Natural stone walkways and patios

Reinforcement inserted in the concrete is often necessary to strengthen it, as well as some expansion joints to prevent cracks. Place joints at least every 3 m (10 ft.) or where the concrete butts against another structure, for example, the house foundation.

Cement manufacturing is energy-intensive. It can be recycled, for example, broken or crushed concrete can be used as surface material. Regular concrete's permeability is low. However, ask about the availability of porous concrete. It uses larger-sized stone particles than regular concrete, making it more permeable and reducing stormwater runoff. Water, instead of remaining on its surface, drains through porous concrete, resulting in savings on repairs and less inconvenience from reduced pooling of water.

Natural stone

Natural stones integrate easily into a natural landscape. This natural, irregular and textured material helps to give a pastoral ambience. Using stone from your region is recommended since transporting these heavy materials over long distances from outside your region consumes energy and creates harmful air emissions.

Depending on how you use them, it is preferable to use the thinnest stone possible in order to save on costs and only use a small amount because these materials are removed from a natural environment. On the other hand, thinner stones have the disadvantage of being less stable. Use stone that is 2.5-5 cm (1-2 in.) thick for walkways and 10-15 cm (4-6 in.) thick for driveways. This thickness for driveways will greatly increase the price of this material. All stones are not equally resistant; it is preferable to check the required thickness with a professional in your region. To reduce tripping hazards, stones should be set to minimize surface irregularities.

Natural stones, installed on a granular base, have joints filled with clean, coarse sand, plants or stone dust (depending on the geographic location). Mortar joints can deteriorate, crack and become loose, especially in regions where ground freezes. Moreover, a patio with concrete joints allows no water infiltration. You can plant groundcover plants between the stones by making the joints slightly wider to enable the vegetation to establish itself and filling the joints with a mix of soil with stone dust or sand. Try Creeping Thyme (*Thymus serpyllu*), Garden Thyme (*Thymus vulgaris*), False Rock Cress (*Aubrieta deltoidea*), Sweet Alyssum (*Lobularia maritima*), Cottage Pink (*Dianthus*) or Dwarf Phlox (*Phlox subulata*). If you remove snow from these surfaces during the winter, the plants may be damaged.

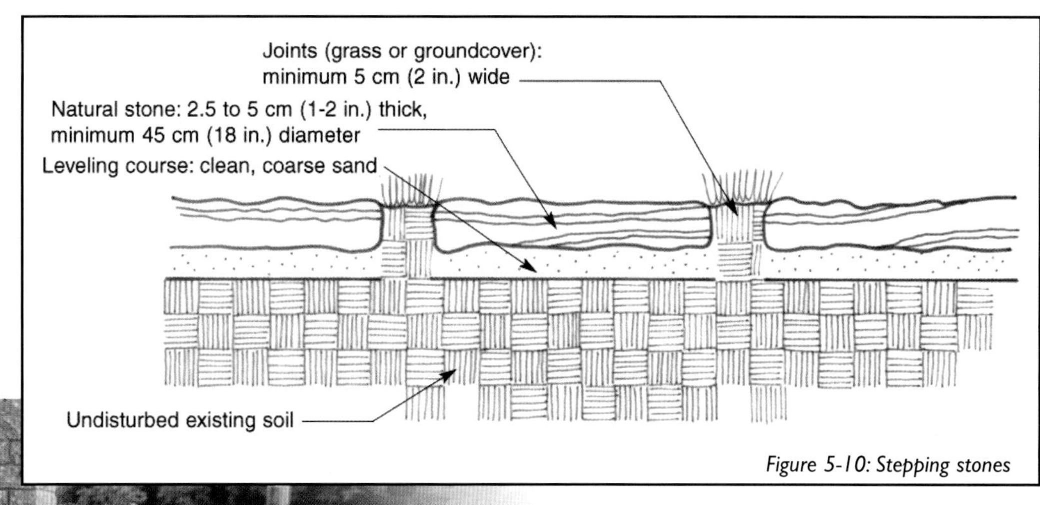

Figure 5-10: Stepping stones

Figure 5-11: Stepping stones provide a firm walking surface but allow permeable surfaces like plants or mulch between the stones.

Stepping stones (Figure 5-10 and 5-11) are useful for lawn or groundcover areas that experience only moderate foot traffic:

1. Use stones that are at least 45 cm (18 in.) in diameter. If stones are too small, they do not provide enough stability and people will walk on the grass more than on the stones.
2. To define the space between the stones, walk slowly on the grass and mark each footstep. The footprint then becomes the centre of the stone. To make the marks, ask the smallest adult in the family to walk along. Lay out the stones to ensure they are spaced correctly.
3. With a knife or edging tool, cut into the soil by following the shape of the stone and remove the grass or other plants where the stone will be placed.
4. Precast concrete slabs can also be used for stepping stones.

Loose aggregate (for example, decorative pebbles, crushed stone, crushed brick)

Loose materials include river washed pebbles, decorative pebbles, pea gravel, stone dust, stone chips, crushed stone, lava rock, crushed brick, crushed concrete and so on. These loose surfaces are inexpensive in comparison with other materials. They are flexible, allowing curving and very irregular layouts and as a result fit well into a natural landscape. They also allow water to seep into the ground, reduce runoff and therefore decrease the amount of water sent to storm sewers, and improve the quality of watercourses.

The problems associated with them are more aesthetic and functional. In some municipalities, their use on driveways is prohibited. The various materials range in price, appearance, texture and comfort as a walking surface (Figures 5-12 to 5-14).

These surfaces require some maintenance. Since they are loose, some material can be removed during snow removal. Install enough material to avoid having to replace it more than every two or three years. And if you want to match the type and colour, keep some aside for future repairs. Weeds often germinate as small amounts of organic matter accumulate in the spaces, for example, from fallen leaves. To cut down on weeds, avoid soil deposits on the surface and mixing loose aggregate with soil underneath. Use a geotextile (filter cloth) between the surface material and the subgrade to reduce mixing of soil from underneath. Removing fallen leaves from these surfaces will also help reduce future weed problems. Some materials can be raked to help remove weeds.

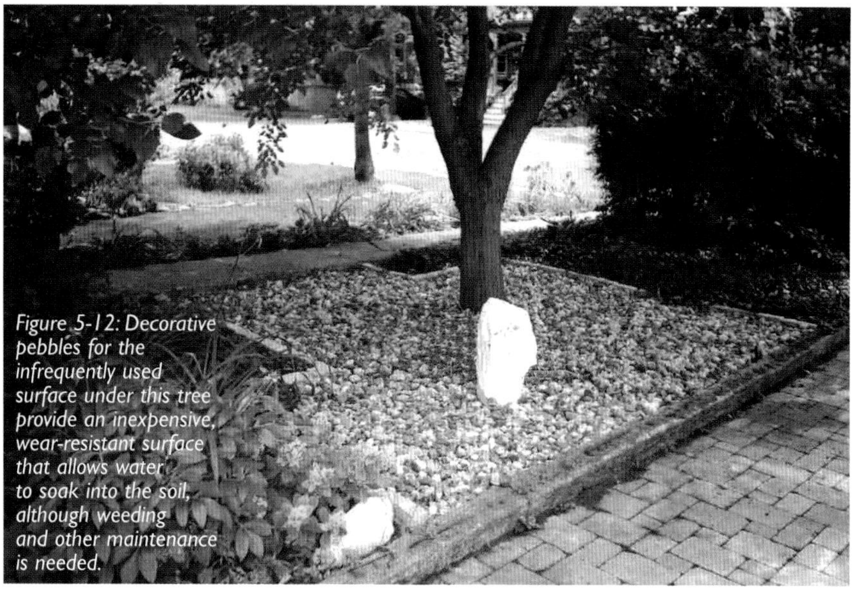

Figure 5-12: Decorative pebbles for the infrequently used surface under this tree provide an inexpensive, wear-resistant surface that allows water to soak into the soil, although weeding and other maintenance is needed.

You can install an edging to delineate these surfaces from adjacent areas, like planting beds, and limit unwanted spreading of the materials. Another way of reducing this problem is to add binders or stabilizers to the aggregate to create a relatively stable hard material.

Loose aggregate, depth depends on type
Base: Crushed stone 0-20 mm (0-3/4 in.) diameter, compacted. Depth varies according to soil type and region; one rule of thumb is 15 cm (6 in.) deep.
Undisturbed existing soil

Figure 5-13: Loose aggregate driveway (for example, decorative pebbles, crushed stone)

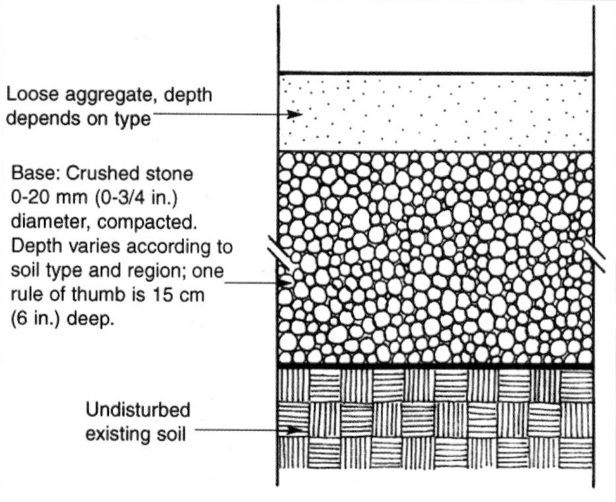

Smooth or rounded loose aggregate surfaces, like decorative pebbles and pea gravel, may not provide a firm footing for people with mobility limitations. Angular materials, like crushed stone or brick may be more slip resistant. Stabilizers or binders increase firmness and stability.

Loose aggregate, depth depends on type
Geotextile to allow drainage and separate the aggregate from the soil underneath
Undisturbed existing soil

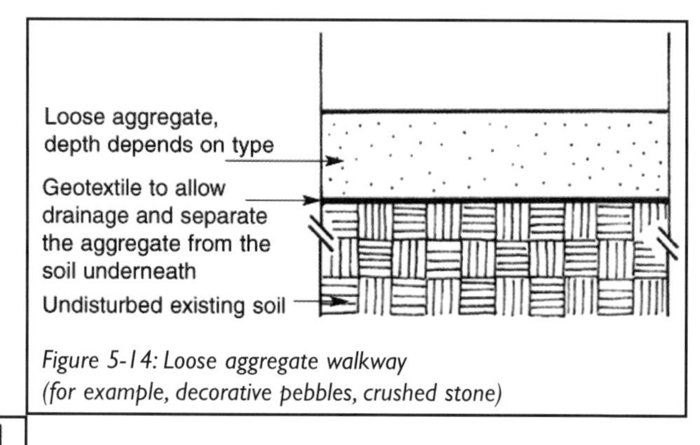

Figure 5-14: Loose aggregate walkway (for example, decorative pebbles, crushed stone)

These materials have environmental benefits. They are permeable and as a result, they let stormwater soak into the soil and reduce runoff, which is better for our water systems. Recycled materials, such as crushed concrete or brick, can also be used.

Reinforced grass rings

A system of manufactured rings for reinforced grass provides a grass surface that can be used for areas where vehicle parking or movement is infrequent.

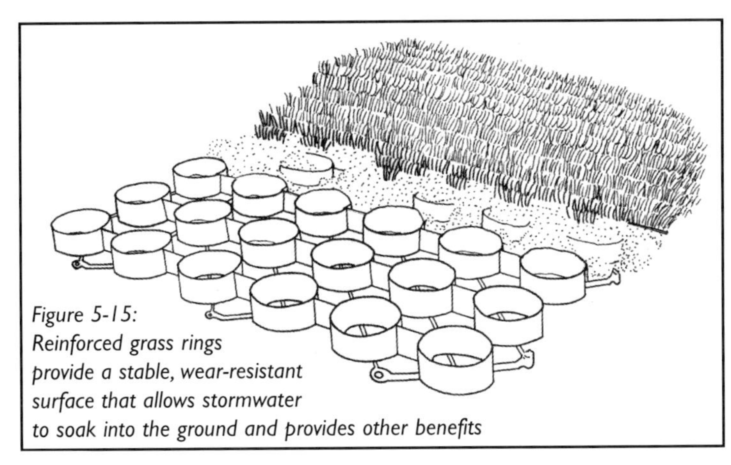
Figure 5-15: Reinforced grass rings provide a stable, wear-resistant surface that allows stormwater to soak into the ground and provides other benefits

For more information on wood chip mulch refer to Chapter 7.

Wood chip mulch creates a soft surface that does not stand up to foot traffic as well as the other materials mentioned. A thick layer of mulch may make access difficult for people with mobility limitations. Apply in low-use areas intended for foot-traffic only, like a pathway through a woodland or wildflower meadow.

When installing reinforced grass rings, fill the cells with soil in order to facilitate the growth of grass. This surface treatment consequently requires more maintenance than paving because it must be mowed and watered. However, this way you can create large areas of green space that would otherwise have been covered over with pavement. It improves water infiltration and reduces runoff. Increasing planted areas also helps reduce hot summer temperatures in urban areas. Favour products that use recycled plastic, since new plastic manufacturing is an energy-intensive process and recycling keeps plastic out of landfill sites.

Wood chips

Wood chips or bark chips/ chunks can be used to create infrequently used paths or sitting areas. They are highly permeable materials. Bark mulch is generally a by-product of milled wood. Wood chips are generally made of tree trimmings, Christmas trees and wood waste, so by recycling these materials, they are kept from being discarded. Mulches also preserve soil moisture and aeration. Once it has degraded, old mulch can be composted and used to enrich planting beds.

Mulch can be displaced by the wind or during snow removal, so clean up may be needed. Weeding will be required as well as replacement of decomposed wood chips.

Edgings

Edgings serve to delineate the areas covered with surfaces such as precast concrete pavers and loose aggregate. There are different types of edging for different purposes: underground types to hold paving in place; and above-ground ones to contain and delineate hard surfaces from adjacent plantings. They have become very popular because, if installed properly, they can reduce the time required for maintenance of paved surfaces.

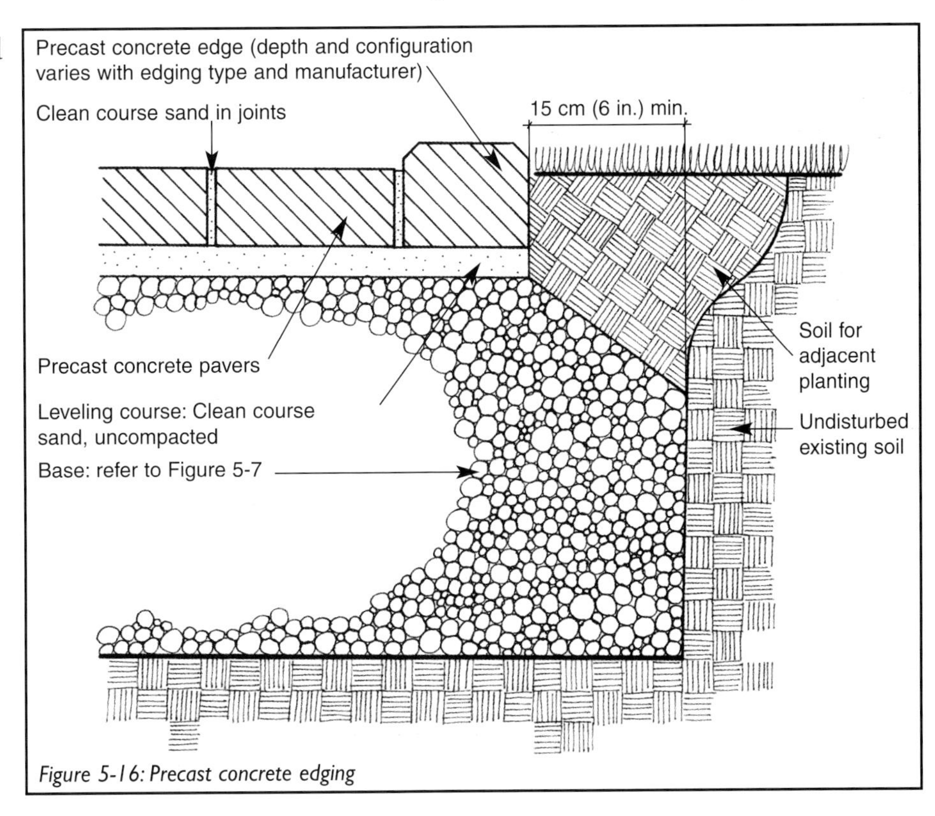

Figure 5-16: Precast concrete edging

Concrete edging

Concrete edgings are available in different widths, heights and lengths to adjust to various situations. They are also available in a variety of styles and colours to enhance the appearance of the design. Install them on the same foundation bed as the paving surface, extending the foundation beyond the paving edge and setting the concrete edging afterwards (Figure 5-16).

If properly installed this is a very durable material that enables a crisp, clearly defined edge to the driveway, patio or walkway. Because it is modular, individual sections can be easily removed and replaced, if repairs are needed. Initial costs are higher than for aluminum or plastic edging.

Plastic or aluminum edging

For precast concrete pavers, plastic or aluminum edge restraint holds the pavers in place and helps prevent them from shifting. Anchors or spikes are used to solidly attach the edge restraint to the foundation where the pavers are placed. This type of edge restraint is lightweight and flexible. Furthermore, it sits underground, making it effective, yet invisible. Refer to Figure 5-7 for installation details.

For loose aggregate materials like decorative pebbles and crushed brick, above-ground plastic or aluminum edging can be used to hold the materials and to define an edge between these surfaces and planted areas (Figure 5-11). Ensure that they are installed properly with anchors or stakes since freeze-thaw action can dislodge them, which would require repairs. Costs are generally lower than for concrete edging. They are moderately durable, depending on whether they are properly installed with anchors. Repair sections that are lifted by freeze-thaw action.

DECKS AND PATIOS

This section provides tips on designing and detailing a deck or patio with a focus on increasing durability, reducing moisture problems, such as decay, and selecting healthy materials, based on the process described in Chapter 2.

1. Identify your needs

Table 5-2 allows you to draw up a list of the activities you anticipate will take place on your deck or patio. You can cut out shapes that correspond to these dimensions and the scale of your plan, and move them around to help you get the right size and layout.

A large patio or deck in a small garden may look inappropriate. Moreover, as described earlier, minimizing impermeable surfaces is better for the environment. To reduce the size of the patio or deck, some activities can be combined or can use the same furnishings.

2. Get to know your site

This aspect was covered in detail in Chapter 2. All information about the immediate site, views or physical elements are applicable to the design of the deck or patio.

Table 5-2: Activity Table

Activity	Number of people	Space required	Location	Sunlight
family meals	4	2.4 x 2.4 m (8 x 8 ft.)	near kitchen	partial shade
coffee nook	2	1.2 x 1.2 m (4 x 4 ft.)	need for privacy	morning sun
sunbathing	2	1.8 x 1.8 m (6 x 6 ft.)	protection from wind	full sun
reading nook	1	1.2 x 1.2 m (4 x 4 ft.)	need for privacy	partial or full shade
children's games	2	1.8 x 1.8 m (6 x 6 ft.)	visible from house	partial or full shade
Spa	2	1.2 x 2.4 m (4 x 8 ft.)	need for privacy	afternoon sun
BBQ space		60 cm x 1.5 m (2 x 5 ft.)	near kitchen near eating location	-
storage		60 cm x 1.8 m (2 x 6 ft.)	accessible	-

3. Locate the best place for your deck or patio

Once you have determined the amount of space and type of location you require, develop your concept plan, using bubbles representing the approximate area required.

During the site inventory/analysis, you identified a number of elements that will affect where the patio or deck will be located. They will enable you to assign each activity more accurately. The most important elements are:

a) Accessibility

Your main deck or patio should be easily accessible if it is used frequently. Its accessibility from the house is particularly important. If it is used for outdoor eating, there will be a great deal of circulation to and from the house, and should have convenient access to the kitchen; the space reserved for the BBQ should also be in close proximity to the kitchen. Universal accessibility is also an important element to be considered.

Designing an accessible deck:

- When designing the size, shape and furniture layout, provide a clear passage at least 90 cm (3 ft.) wide.
- Provide a deck level at the door that is no more than 1.9 cm (¾ in.) below the inside floor level. The threshold should be sloped to the exterior.
- For access from the deck to the yard, follow the tips on wheelchair-accessible paths discussed earlier in this chapter. Contact your municipality for code requirements on ramps.
- Provide cooking and food preparation surfaces at a height of 86 cm (34 in.).

For more information on barrier-free design, look for these CMHC publications: *Housing for Persons with Disabilities,* and *Housing Choices for Canadians with Disabilities.*

Figure 5-17: This accessible deck has a ramp with a 10 cm (4 in.) high border and gentle slope.

b) The views

On your site inventory/analysis plan, the views you want to preserve and/or hide have already been identified. They will be useful for planning the deck or patio.

Since this is often the most frequently used area in the yard, and also a place for sitting and relaxing, you may want to benefit from the best possible view. Locate the deck or patio to take advantage of an attractive panorama over your garden or even over the surrounding environment. There may be, for example, a beautiful plant bed at the back of the yard or a large tree in the neighbour's yard. If there is no outstanding feature, you can create one yourself.

Conversely, some views are unpleasant and you may want to block them. Particularly on small properties, neighbours can be quite close and privacy can be a problem. You may have to isolate yourself from the decks of three other neighbours and the windows that can overlook your yard. The positioning of the deck or patio, or keeping the deck low to the ground and using plants or structures to screen views, can help create privacy and intimacy.

c) Sunlight and wind

Sunlight is also a factor to consider. As we indicated in Table 5-2, the activities planned on the deck or patio are dependent on a particular level of sunlight. For example, although sunbathing requires a maximum of sun, eating meals may be more pleasant in partial shade.

Exposure to wind is also a factor you should look at. A deck or patio on an exposed lot may become unpleasant if the wind is too strong. Eddies and wind tunnels are phenomena that occur on a small scale and can make certain areas uncomfortable. Planted or built screens may help remedy the situation but it's better to prevent it by finding the right location.

d) Type of Soil

The type of soil may also be a limitation. A deck or patio's foundations may be laid out in practically any type of soil. If installing a patio in clay soil, the gravel base should be deeper than for other soils. For a patio, look for areas without steep slopes and where there is sufficient soil depth, for example where the bedrock is not too close to the surface.

4. Check the regulations

Early in the planning phase, call your municipality to find out about local codes and bylaws. You may need a permit and there may be certain requirements for guard rails, size and spacing of beams and joists, size of steps, etc. You may be required to submit a landscape plan with your building permit application.

Check also the required setbacks in your municipality; these distances regulate how far structures must be from the lot line, and may limit your choices when deciding where to locate your deck.

5. Design the layout

First think about the height. The closer the deck is to the ground, the easier it is to have privacy. Even so, you may still need to install fences or plant screens to get the privacy you need.

If your deck is installed at the level of the finished floor of the house, the indoor link is more direct. A two-levelled deck may be preferable. The first level, built close to the house, will allow you to go outside easily. The lower level could be larger and provide more privacy.

Locate stairs and landings to give the best access to your house and other parts of your property. Verify that the deck does not conflict with some of the home's existing elements such as a basement window or a utility hook-up. Try to design around those features. For example, if the deck is raised, you could create an opening to allow the sun to filter through to a window.

The easiest solution is to build your patio or deck in a square or rectangular shape. The simple shape requires a minimum of work. Many publications are filled with inspiring ideas that can be adapted to your site. Examining decks or patios in your neighbourhood, especially those of homes similar to yours, you can see which designs inspire you the most. It is the best way to get an idea of how various designs and colours blend in reality.

After making different sketches trying different configurations on your plan, draw all the elements and furnishings to verify if they fit and that there is enough space for movement and other activities. Or you can reproduce the layout on the ground using string and pegs. Then you can move the furniture around it to check the resulting dimensions and views.

Figure 5-18: This deck has been planned around an existing tree.

Design your deck around existing trees and frame an opening in the joists wide enough to accommodate sufficiently the tree's mature trunk diameter. (Figure 5-18)

6. Work out the technical approach and choose the materials

For patios, refer to the comparison of materials in Table 5-1 earlier in this chapter. Table 5-3 compares various deck-building materials.

Pressure-treated wood

This is lumber, for example, pine that is treated with chemical preservatives that are forced into the wood under pressure. Wood treated with chromated copper arsenate or CCA has been commonly used for decks for over 30 years. However, there are concerns that the preservative, which contains arsenic, may leach out over time, especially when in contact with water or soil and may result in unacceptable exposure for people. While the Pest Management Regulatory Agency (PMRA) has deemed it to be safe in the past, at the time of publication of this book, the agency was re-evaluating the risk to the public and the environment. Refer to the PMRA Web site www.hc-sc.gc.ca/pmra-arla for more information. As of December 31, 2003, manufacturers no longer use CCA to treat wood for non-industrial uses such as decks, play structures, picnic tables, landscape timbers and fences, although remaining stocks of CCA treated wood can still be sold in stores and used in Canada. Wood preservative alternatives registered for use in Canada are alkaline copper quaternary (ACQ) and copper azole.

If you have concerns about existing CCA-treated wood on your property, limited studies suggest that the application of certain penetrating coatings (such as oil-based, semi-transparent stains) on a regular basis for example, once a year or every other year depending upon wear and weathering, may reduce the migration of wood preservative chemicals from CCA-treated wood. "Film-forming" or non-penetrating finishes, like paint, can peel and flake, which may reduce durability and potentially increase exposure to the preservative in the wood.

To reduce potential risks, consider alternatives, such as cedar, plastic lumber (plastic-wood composite) or building a patio. To minimize contact, you can limit the use of pressure-treated wood to framing elements such as posts, beams and joists, and use cedar or plastic wood for decking and features above-deck level.

Basic precautions for pressure treated wood

If you use pressure treated wood:

- Take basic precautions when you work with it:
  - Wear a dust mask, protective goggles, gloves and a long-sleeved shirt when cutting, sanding or drilling this wood in order to prevent sawdust from coming in contact with your eyes or skin. Wear gloves and long sleeves when handling it.
  - Wash your clothes separately afterwards.
  - Wash hands and other exposed skin after contact and before eating, drinking or smoking.
  - Cut or otherwise work with it outdoors only.
  - After construction, all end cuts, sawdust and construction debris should be cleaned up and disposed of in accordance with local regulations.
- Do not use it as a surface on which food will be prepared. Food should not come into contact with any treated wood.
- Do not use sawdust or chips from pressure treated wood in your garden.
- Do not burn leftover wood in a fireplace or a wood stove.

Cedar contains natural preservatives, which can cause irritation for some people, especially when they are exposed to the sawdust. You can minimize exposure to these elements by wearing a mask, eye protection and protective clothing when cutting.

Table 5-3: Choice of deck material

Surface material	Durability	Initial cost*	Characteristics	Repair and maintenance	Environmental considerations**
cedar: western red cedar and eastern white cedar	average to long	high	- it contains natural preservatives. - structurally, not as strong as spruce-pine-fir (SPF) woods. Install joists closer together or use solely for visible parts of the deck such as decking, railings and stairs. - colour greys over time. Red cedar has a more reddish brown colour than white cedar.	- easy to work with - cedar may be left natural or finishes can be applied to preserve its colour. - soft, easily marked wood	- recyclable - reuse or recycle cutting waste - natural decay resistance, no added preservatives
pressure treated wood (spruce, pine, fir - SPF) and hemlock, fir (hem-fir)	long	average	- has been chemically treated with preservatives. See page 79 for more details on the preservative treatments. - gives wood a green colour. - pine is used most frequently. - apply preservative to all end cuts.	- easy to work with but follow basic precautions listed on page 79 above for treated wood	- non-recyclable- dispose of cutting waste. See note on page 79 on treated wood - refer to note on page 79 for safety, health and environmental issues of treated wood.
SPF and hem-fir (not pressure treated)	short	low	- higher maintenance - durability is low - need to regularly apply penetrating finishes, like stains and replace decayed sections - for above-ground only	- regularly apply finishes to exposed surfaces (top, bottom, sides, ends) - replace decayed sections more frequently than other options	- low durability - needs more frequent replacement - requires regular application of finishes
plastic lumber (plastic-wood compo-site)	long	highest	- made of wood fibre, such as sawdust, and plastic, recycled or virgin - used for decking, railings and siding. Products with structural strength to be used as posts, beams or joists not yet available. - relatively new. Products vary widely depending on manufacturer, e.g. durability - heavier than wood but some products are hollow. - all plastic and vinyl deck materials also available (also a maintenance-saving option)	- low-maintenance - does not need to be preserved, painted or stained	- recycled product - diverts both wood and plastic waste from landfill but some use virgin plastic, which has energy consuming manufacturing process - does not contain preservatives with arsenic or other heavy metals - recyclable

*More durable materials can be less costly in the long run, since they have lower maintenance and replacement costs

**Transportation energy is reduced by choosing materials harvested and manufactured locally. Environmental impacts of these materials depend on raw material harvesting and regeneration practices.

For more details on decks, including foundations, beams, joists, lumber sizes and more, refer to the CMHC publication *Healthy Housing Renovation Planner.*

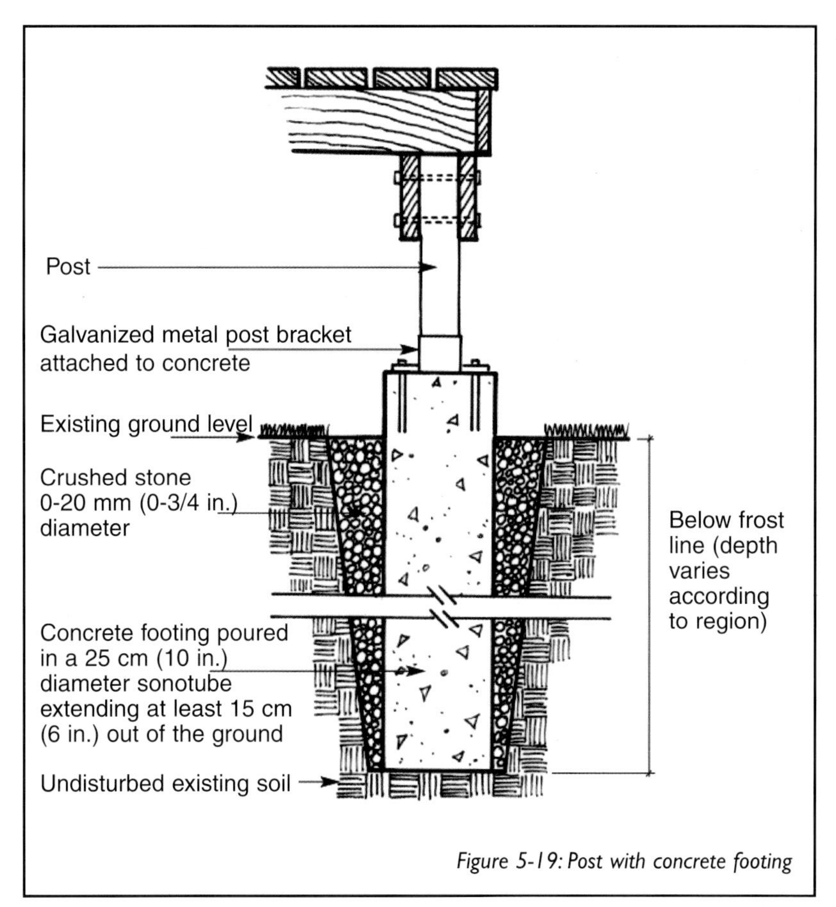

Figure 5-19: Post with concrete footing

Among the wood options, only cedar has natural properties that protect it against decay and insect damage without requiring pressure treatment with chemical preservatives or regular applications of penetrating finishes. For more information on pressure-treated wood, refer to page 79. Plastic lumber is a relatively new, low-maintenance decking material that can be made of recycled materials.

Tips for enhancing deck durability and protecting it from moisture

Figures 5-19 and 5-20 show an example of deck construction. Different types of assemblies can be used. There are many publications available on deck design and construction. The following tips focus on design and detailing that help make decks more durable by protecting them from moisture and decay.

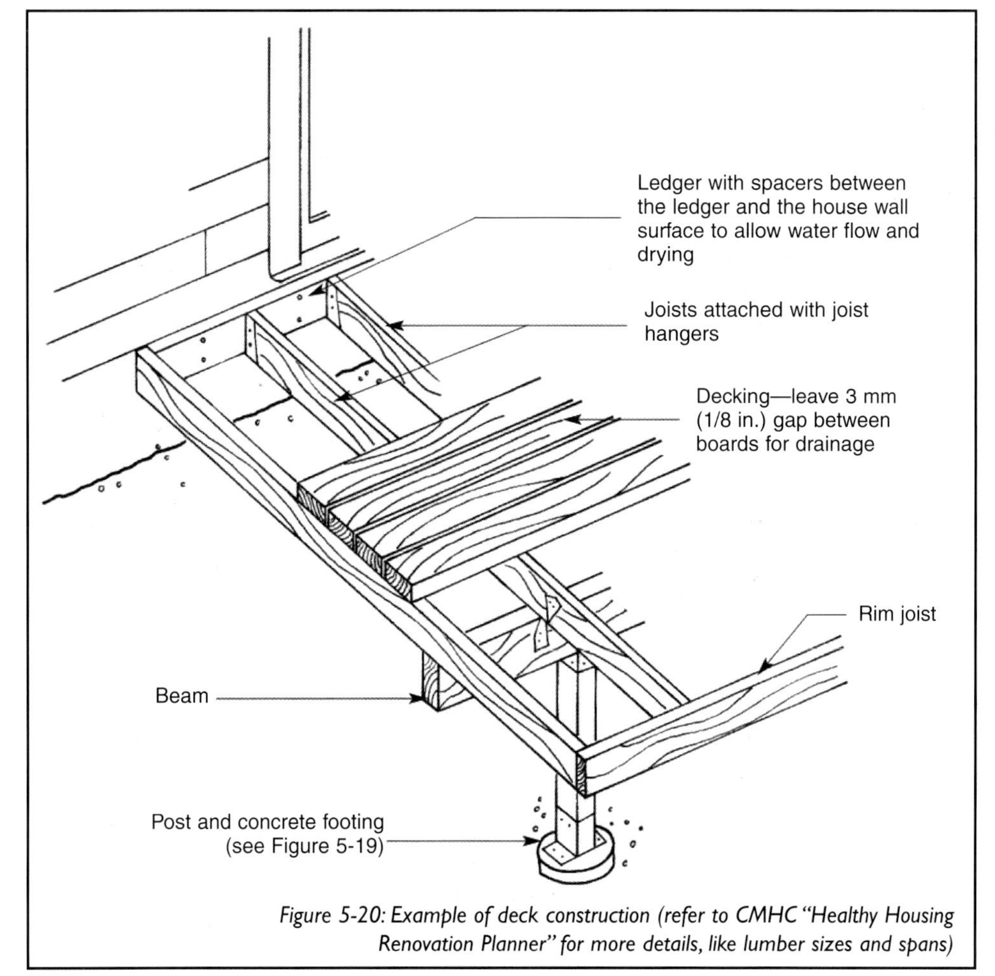

Figure 5-20: Example of deck construction (refer to CMHC "Healthy Housing Renovation Planner" for more details, like lumber sizes and spans)

Foundation:

Figure 5-19 illustrates a concrete footing. Projecting the footings above the surface of the ground so that the wooden posts do not touch the soil is a good way to avoid premature decay, making the deck more durable. Sometimes this foundation can be made of a simple concrete pad placed under each post for low decks in areas where the ground is stable and not likely to heave with the frost. This method is simple and less expensive.

Ledger:

Most decks are attached to the house using a long board called a header or ledger. This board is attached to the wall using lag screws or bolts fastened into the floor frame or the wall of the house. To reduce moisture problems, create airspace between the house and the ledger by using spacers, such as washers on the screw or bolt between the house and the ledger. If you are attaching the ledger directly over the house's siding, you can nail shims, for example 1 cm (3/8 in.) thick preserved wood or cedar, to the back of the ledger every 40 cm (16 in.) to create an airspace between the siding and the ledger. The ledger is often fastened to the basement concrete, using expansion bolts or concrete anchor bolts. Joists can be attached to ledgers using metal joist hangers.

Beams and joists:

The beams can be set directly on top of the posts, or bolted to their sides. Joists sit on top of the beam or at the same level. Small pieces of wood, called blocking or cross-bridging, can be nailed between the joists to provide greater strength. For information on beam sizes, joist spans, decking selection and joist spacing, refer to the chapter on Decks in CMHC's *Healthy Housing Renovation Planner*.

Use wood efficiently

When buying wood, you can avoid waste by buying sizes that will need the least cutting and produce the fewest short cut-off pieces. It helps to make your deck size conform to standard lengths of lumber, which come in incremental lengths of 61 cm (2 ft.); a small change in the dimensions on your plan can save money and wasted wood.

Decking:

A 3 mm (1/8 in.) space should be left between the deck boards to allow rain to drain through. The deck surface should be at least 19 mm (¾ in.) below the door threshold to prevent rainwater from entering the house. The deck surface should also slope a minimum of 1 per cent away from the house, to ensure its drainage.

Fasteners:

There are a wide range of metal fasteners and straps, like joist hangers, that can simplify the work. All these hardware items should be made of hot-dipped galvanized metal or stainless steel, otherwise they will quickly rust, stain and lose their structural strength.

Edges:

All wood edges cut during installation should be treated with a paint-on preservative or stain. Before applying the preservative, refer to the basic precautions on treated wood stated earlier in this chapter. The most vulnerable spot is the ends of the boards; these can decay easily, especially when they are nailed up against another board where moisture can be trapped. Using metal joist hangers, which provide a small space at the end of the joist, can help prevent this type of decay.

Building a deck is a project that many homeowners can carry out themselves. However, because decks require carpentry skills, special tools and lifting, it may be best to hire a professional.

Under the deck:

Ensure that there is air circulation under the deck so that the wood does not stay damp, for example, by using lattice to close in the sides rather than solid wood. Under the deck, cover the ground with a surface material that will allow water drainage but limit the spread of weeds, such as crushed stone. To allow water drainage, the ground surface should have at least a 5 per cent slope away from the house but a 10 per cent slope for the first 2 metres is preferred to ensure the correct slope after the soil settles and compacts. More details on drainage around the house are provided in Chapter 4. A clutter of leaves left lying on the deck surface can also trap moisture against the wood, so it pays to keep the surface clean and free from debris.

Railings:

Most decks need a railing to keep people from falling off. The National Building Code (NBC) requires a railing or guard for decks over 61 cm (2 ft.) off the ground. They must be at least 91 cm (3 ft.) high for decks up to 1.8 m (6 ft.) above the ground. For higher decks, the railing should be 1.06 m (3½ ft.) high. The space between the slats should be less than 10 cm (4 in.) for the safety of children. The NBC requires that the design and construction of railings prevent climbing. These railings may be supported by the deck posts or by separate posts fastened to the frame. Stairways with more than three steps should have a railing 81 cm (2 ft. 8 in.) high on at least one side. For railing requirements, check your municipality.

Finishes and maintenance:

You can simply leave cedar and pressure treated wood unfinished, and after a few years it will take on a grey hue. However, most people want to maintain the initial appearance of the wood; they apply a finish to blend with their house materials that will provide a certain amount of protection against rain and sun.

Among the products available, the most common are penetrating finishes like semi-transparent deck stains, sealer or water repellents. These products penetrate into the wood and often contain a wax that helps repel water, prevent mildew and provide protection from the sun's ultraviolet (UV) rays. However, they have to be reapplied on a regular basis. Penetrating oil stains are the only finishes that should be used on pressure treated wood. Film-forming finishes like paint and solid or opaque stains do not penetrate the wood. They can blister, peel and be scratched, allowing water to be trapped next to the wood, where it can lead to decay.

Most of the stains and finishes made for decks are oil-based. Just like oil-based paints or alkyd paints, they contain solvents that give off irritating fumes as they dry. Just as for paints, water-based finishes (usually called "latex") are a healthier choice. To better protect yourself from fumes, especially if you are sensitive to the odours, wear a mask during the application and wait until the surface is really dry before using it.

GREEN ROOFS AND ROOFTOP GARDENS

Green roofs are roof surfaces used to grow plants. They can be accessible to people as gardens or terraces, making effective use of limited space

Figure 5-21: This green roof is on the Merchandise Lofts Building in downtown Toronto. For more information on this project and another green roof at the Waterfall Building in Vancouver, refer to CMHC's Web site www.cmhc.ca
Photo: Terry McGlade, Perennial Gardens Corporation

Tips for roof decks or balconies:

- To help reduce wind exposure for the plants, group pots or planters closely and locate them near walls. Some people anchor their planters to the floor or wall with wire. Since roof gardens can be windy, they may not be suitable for hanging baskets.
- These exposed locations can also cause dry conditions. A faucet on your roof deck or balcony will make watering easier. Use a few large containers rather than many small ones and choose plants that tolerate dry conditions.
- Use a light-weight growing mix designed for containers.
- Since roof decks and balconies are normally small spaces, ensure that the mature height and spread of plants is not too big for the space. Vines trained on trellises or other vertical structures are good options in tight spots.

on small lots and apartments. They can also be used just to grow plants and accessed only for limited maintenance. In this case, the plants, light-weight growing medium, drainage layer and waterproof/root-repellent membrane are an integral part of the roofing system. They reduce stormwater runoff from roofs, filter the air and help cool cities in the summer since plants absorb solar energy and use it to grow instead of re-radiating it as heat. They also provide insulation and help reduce heating and cooling energy costs in your home. For example, the National Research Council found that a green roof can reduce heat flow through the roof by more than 75 per cent. For more information, refer to the Green Roofs for Healthy Cities Web site at www.greenroofs.ca/grhcc, which includes technical information on how to install a green roof.

Planters

Planters are particularly useful on decks, balconies and roof surfaces. Decide what type of plant you want and design the size of the box accordingly. A rule of thumb for soil depths is

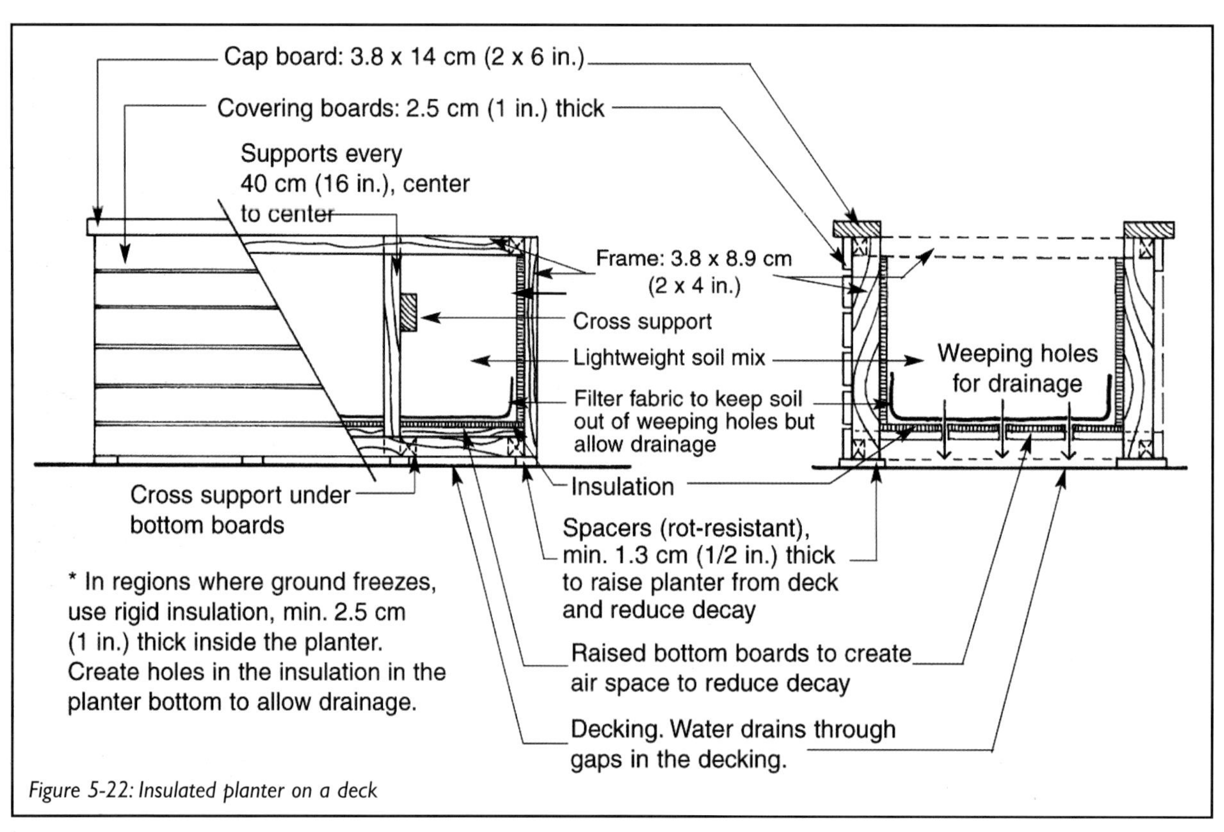

Figure 5-22: Insulated planter on a deck

Figure 5-23: Raised planters are ideal for people who have difficulty bending over to garden. Photo: courtesy of North Renfrew Long Term Care Services Inc., Deep River, Ontario

to provide at least 30 cm (1 ft.) of soil for perennials; 60 cm (2 ft.) for shrubs and 1 metre (3 ft. 3 in.) of soil for trees.

Soil saturated with water expands when it freezes, which tends to crack or make the walls of a rigid container bulge. In regions where ground freezes, the roots of perennials, shrubs and trees in planter boxes should be protected against freezing and thawing. Install rigid insulation on the inside walls to

The Earthwise Garden: Delta Recycling Society, Delta, British Columbia

The design of the imaginative Earthwise Garden shows that using reused and recycled building materials in the garden can be a fun and creative process! The Earthwise Garden was built in 1995 as a public demonstration garden illustrating the linkages between recycling and ecological sustainability. Using recycled material throughout the garden not only kept construction costs down, but also provided the satisfaction of knowing that materials were diverted from the landfill and that the environmental impacts of harvesting or manufacturing new materials were avoided. Here are some creative ways in which recycled and reused materials were utilized in the garden:

- The roof of the gazebo is made from reused tires.
- The gazebo, entry arbour and bridges are made from dimensioned lumber reclaimed from old hydro or telephone poles.
- Fences are constructed out of sections of old telephone poles with discarded cable strung between each pole.
- Recycled concrete pieces are used in the patio (see photo). The patio resembles a natural stone patio, but costs significantly less.
- Pathways are covered with recycled crushed concrete.
- Soaker hoses are made from recycled tires and rain barrels.
- Compost bins are made from old pallets.
- Branches and twigs pruned from the garden are used for plant supports, arbours and benches.
- Community compost obtained from the municipal waste management facility was used for soil amendment when the garden was built, and annually for mulch and top-dressing.

Community volunteers involved in the ongoing maintenance of the garden are always on the look-out for unusual items to include in the garden. Birdhouses have been made from reused fence boards (see photo). It often takes a bit of creativity and research to find recycled or reused materials. Often the materials are available for free or low costs, although delivery costs may apply. Children and adults visiting the garden are amazed that they can use something old for a new purpose.

Photos: Delta Recycling Society

A small backyard paradise: Chase-Cross Garden, Halifax

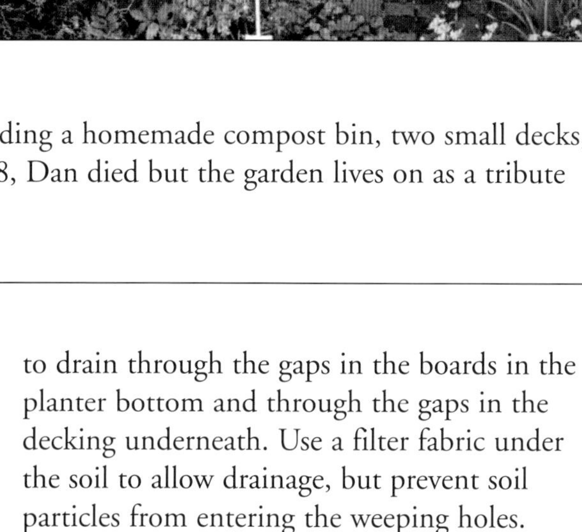

When Tim Cross and Dan Chase began work on their backyard in Halifax in 1984, it contained rubble from renovations and the ground was compacted from years of parking and neglect. They began by reclaiming the fence pieces and protecting the space from traffic. They broke up the ground with a spade and rake and sifted it to remove rocks and debris, like coal and glass. They gradually expanded a perennial garden and created a patio over what used to be a marginally successful lawn. They collected plants from family and friends, mostly perennials, and eventually added four small trees; one for each season. The garden fence and a delightful pergola completed the picture. The new fence transformed the garden into another room and provided a great backdrop for vines.

The small garden is 9 m (30 ft.) by 5.5 m (18 ft.) with a 7 m (24 ft.) long brick path and narrow bed along one side of the property. They have made the most of a small space, including a homemade compost bin, two small decks, a pergola and space to hang a hammock. In 1998, Dan died but the garden lives on as a tribute to his memory.
Photo: Tim Cross

protect them from freeze-thaw action and reduce moisture penetration from the soil to the wood, which will help extend the life of your planter. In regions that do not experience freezing, you can use an impermeable liner instead to protect the wood from soil moisture. Slit the liner in the bottom of the planter above the drainage holes.

Consider the strength of the walls of large planters to ensure they can withstand the internal pressure caused by soil and water. By installing weeping holes through the insulation in the bottom of the box, you will allow the water to drain through the gaps in the boards in the planter bottom and through the gaps in the decking underneath. Use a filter fabric under the soil to allow drainage, but prevent soil particles from entering the weeping holes.

For more information on materials, including wood, fasteners and finishes, refer to the section on decks. Provide air space between the bottom of the planter and the decking by raising the planter bottom as shown in Figure 5-22. You can use thin decay-resistant spacers underneath the outside frame to separate it from the deck.

PLANTING DESIGN

Chapter 6

Plants perform a variety of functions that will help you meet your needs. They beautify your home and community, block views and winds, create shade, and a host of other benefits. They also let stormwater soak into the ground, reduce runoff and erosion, create habitat and cool city temperatures in the hot summer. Their selection, arrangement and maintenance affect the use of water, energy, fertilizer and pesticides. This chapter gives several ideas for considering the environment in your landscape design, while meeting your needs and helping you save time and money on maintenance. It also describes various functions plants perform and approaches to planting design that are inspired by nature. Technical advice on how to install and maintain the plants in your design is provided in Chapter 7.

In Chapter 1, we discussed the advantages of working with nature. Natural plant communities in your region are adapted to the natural cycles and physical conditions they find themselves in. As a result, nobody needs to water them, fertilize them, apply pesticides, mow, rake or trim them. All the elements are interconnected, including the plants, animals, soil, wind, sun, shade, climate and moisture. Even for small yards, letting your design be inspired by nature will not only save you time and money on maintenance, but is also better for the environment. For example, planting a woodland shade garden, wildflower meadow or prairie garden on your property, depending on which is more suited to your location and property conditions, can reduce your use of water, energy, fertilizer and pesticides. You can also celebrate a small piece of the unique diversity of the flora and fauna of your region.

When doing your planting design, remember the process described in Chapter 2. In summary: Know your needs, get to know your site and find the best location for each feature. A thorough site analysis is critical because each plant should be suited to the specific site conditions it finds itself in, such as space, sunlight, soil and moisture conditions. Next, check regulations, do the preliminary layout and choose the plants. You can reproduce the general layout on the ground using string and pegs or a garden hose to more realistically see if the layout works. Finally, work out the technical approach, estimate costs in light of your budget, and refine the plan.

For a planting design that meets your needs and works with nature:

1. Preserve natural features and existing plants
2. Know your needs and the role plants can play in your design
3. Get to know your site conditions and choose suitable plants
4. Choose low-maintenance planting options that work with nature, like xeriscapes, woodland shade gardens or wildflower meadows

Healthy, mature trees can increase a property's value significantly. Professionals such as arborists and landscape architects can help you assess the value of your trees.

1. Preserve Natural Features and Existing Plants

Even with the best intentions, it is difficult to recreate a natural environment, such as a woodland, once it has been destroyed. Unfortunately, when most sites are developed, vegetation is removed as the site is regraded and backfilled. A few trees might be spared since they increase the value of the properties, yet they can die in the following years due to damage suffered during construction, or changes made to the grading, drainage and sunlight conditions around the trees.

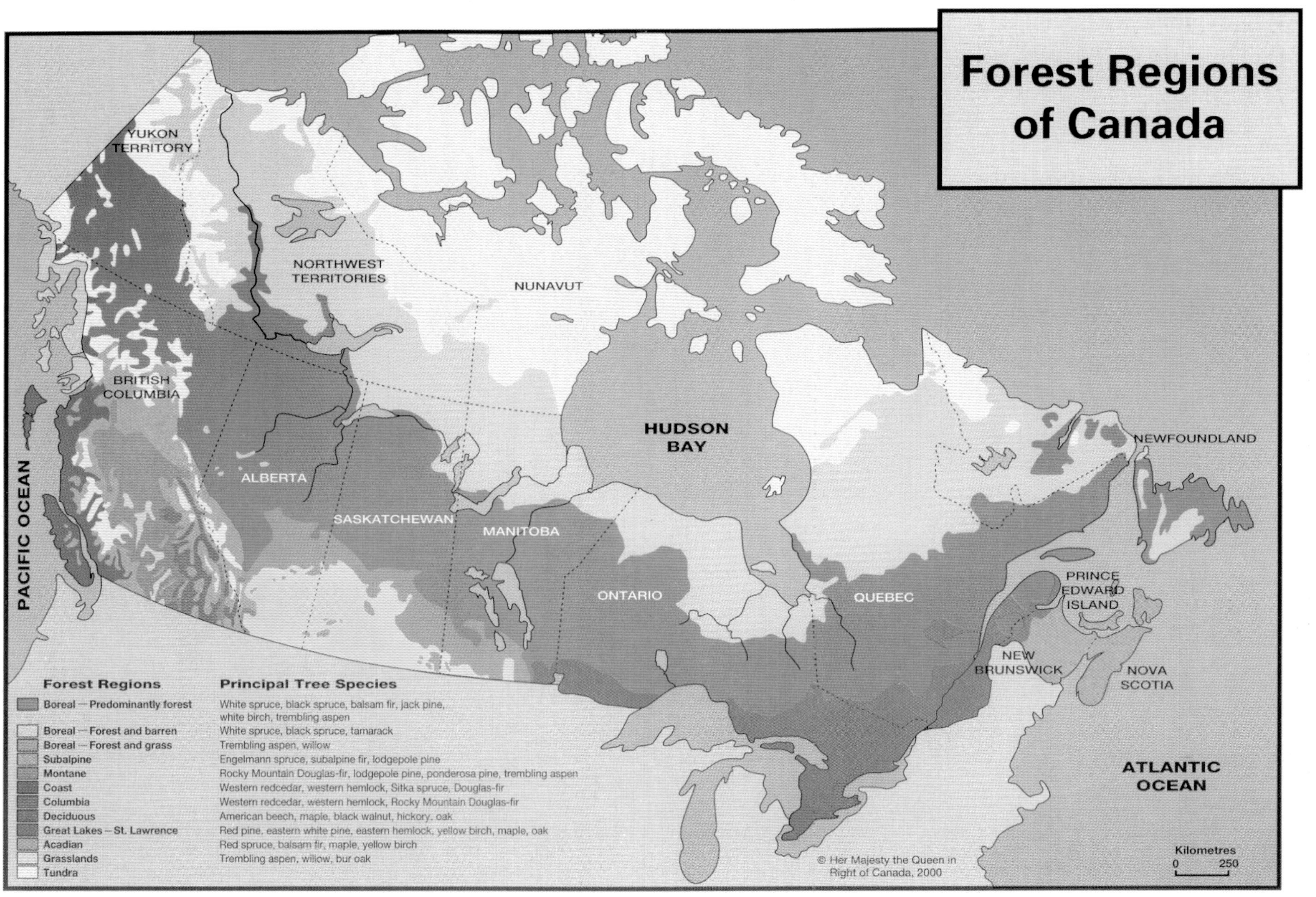

Figure 6-1: Forest Regions of Canada
Reproduced with the permission of Natural Resources Canada, Canadian Forest Service

A house in the woods: The FlexHouse of Saint-Nicolas, Quebec

Photo: Jean-Marie Lavoie

This house is nestled in a forest near Québec City. Through careful site planning and effective protection of vegetation during construction, a minimal number of trees were removed. This project protects the environment in many ways:

- The streets in the neighbourhood are very narrow which helped preserve the forest. The house is located close to the street and the driveway is short and narrow. Had the house been located further into the site, a much longer driveway would have been required, resulting in more tree removal.
- There are no lawn areas on the property. As a result, no trees were removed for this purpose. The understorey consists of low-maintenance groundcovers and understorey plants found in the boreal forest.
- During construction, vegetation was protected and existing trees remained very close to the house. Construction equipment, material storage and grading were prohibited in the areas intended for preservation. Ecological servitudes were drawn up to ensure the forest survival.
- The house is built on a concrete slab, rather than over a basement; therefore it wasn't necessary to dig a deep foundation. As a result, fewer roots were cut and more trees were preserved near the house. No underground drainage systems were used. Even the roads are surface drained in ditches. Water stays on the site to preserve the water table.
- Since the house is a three-storey structure, it takes up less space on the lot than it would have if it were a lower, more spread-out building, like a bungalow. Again, this helped to minimize the removal of trees needed to make space for the house.

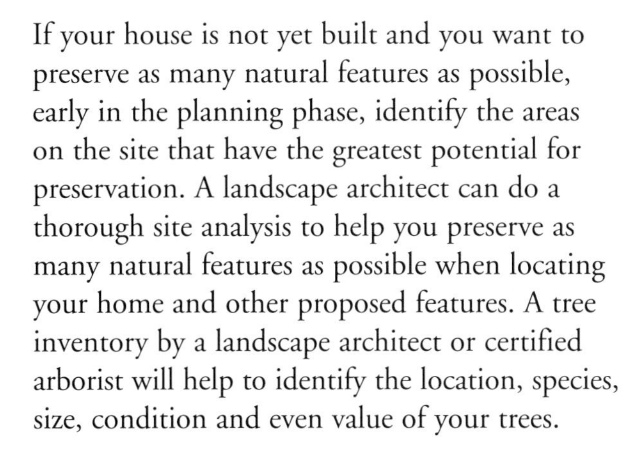

If your house is not yet built and you want to preserve as many natural features as possible, early in the planning phase, identify the areas on the site that have the greatest potential for preservation. A landscape architect can do a thorough site analysis to help you preserve as many natural features as possible when locating your home and other proposed features. A tree inventory by a landscape architect or certified arborist will help to identify the location, species, size, condition and even value of your trees.

Check with your municipality to determine if there are regulations concerning tree protection that would affect you. Some cities have bylaws protecting trees on private property. Also contact your municipality for regulations that may apply to projects that affect natural heritage features such as environmentally sensitive areas, wetlands and other important ecological features. More details on regulations are provided in Chapter 2.

As it is often impossible to retain all natural features of a site when you are building a home, you can divide your property into different zones:

Area 1—Landscaped area (intense construction activity). This includes the house, driveway and patios. In this zone, most species will need to be removed but may be replanted elsewhere on the site.
Area 2—Transition zone (area to be restored). This is the space reserved for vehicles, material and equipment during construction. When work is complete, it will be replanted.
Area 3—Protected area (for leaving intact). Maintain existing grades and prohibit all construction work, equipment and material storage in this area during construction. Install a protective fence at the edge of the transition zone.

Plan the location of the house, driveway and utilities to preserve as many desirable trees and other plants as possible. For example, if your lot is wooded and you must remove trees to locate a driveway, build your home close to the road rather than deeper into the site, to keep the driveway as short as possible. Also try to keep the driveway as narrow as possible and design the layout so that it leaves desirable trees intact (see figure 5-1). A house of two storeys or more that takes up less space on the ground than a bungalow will help you preserve more trees. Ensure that trees to be preserved and their fenced protective zones are clearly marked on the property and on the site plan.

Deciding whether a tree will survive after construction due to its location or state of health is difficult. For example, a tree's survival will be affected by changes to its exposure to wind and sun when attempting to save a single tree while removing others clustered around it. There may not be enough roots left once the building, driveway and utilities are in place, even if these elements are located as sensitively as possible. Consider hiring a professional, such as a certified arborist or landscape architect for assistance.

Figure 6-2: Tree roots are shallower and wider reaching than you might think, extending well beyond the drip line of the tree.

Establish a clearly defined protective zone with a temporary but sturdy fence, about 1.4 m (4.5 ft.) above the ground, around trees you want to preserve. For best results, the protective zone should extend beyond the drip line of the tree (the ends of the outer branches) as tree roots normally spread out well beyond the drip line (Figure 6-2 and 6-3). One approach is to extend it to one foot of radius for each inch of trunk diameter, as measured at breast height. Any grading, excavation, movement of construction

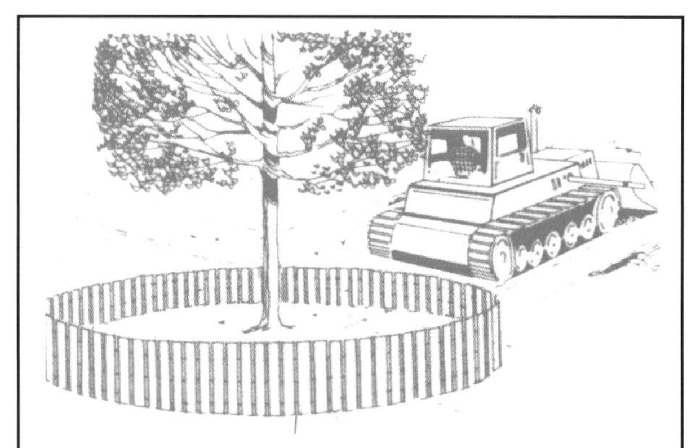

Figure 6-3: Establish a clearly defined protective zone with a temporary but sturdy fence around the tree, preferably well beyond the drip line of the tree to prohibit access by heavy equipment, grade change and other construction activities like storage of building materials.

Properties adjacent to a woodland or other natural area:

If you have a woodland or other natural areas, like a creek or wetland, adjacent to your property, leave native plant materials untouched, including the forest understorey and fallen or dead branches and leaves. They provide habitat and enrich the soil, so if you want to "clean up," limit your activities to removing only human-generated litter, like bottles and cans. Structures, unauthorized paths or other alterations on these lands can contribute to the deterioration of these natural areas. Dumping and piling materials like yard waste in these areas can also cause damage. Yard waste containing invasive species can lead to their dispersal and rapid spread into nearby natural areas, where they may crowd out native species. Use non-invasive plants on your property. If you live beside a body of water, refer to the section on Living by Water in Chapter 4.

Figure 6-4

equipment and storage of construction materials or dumping of waste should be prohibited within this zone. The posting of tree protection signs on the fence may also help to notify other contractors on the site of the required tree protection zones. Refer to Chapter 3 for details on preserving trees in areas where you are making grade changes. Shrubs, groundcovers and other plants within the protective zone should also be preserved. Shrubs, perennials and small trees that can't be kept in their original location may be relocated. For plants that may be difficult to transplant, consider collecting seeds or cuttings.

Even on the small scale of your lot, ask yourself if your decisions about the vegetation on your property will affect local wildlife habitat. Tips for preserving and restoring wildlife habitat, including corridors are provided in Chapter 7. You can choose to give part of your property over to nature by preserving the parts that link with other natural areas. For example, if a wooded area runs along the back boundary of the lot, leave that part of property intact to expand the woodland as much as possible. (Figure 6-4) By discussing it with your neighbours, you may be able to preserve or create a much larger wildlife habitat that runs from one lot to another.

2. Know Your Needs and the Roles Plants Can Play in Your Design

Plants serve many purposes, including:

a) Pleasing visual qualities

An infinite number of visual effects can be achieved in your planting design. Plants should be selected and arranged to take advantage of their colours, textures, shapes and sizes. In terms of visual effects, they can:

- define space or pathways
- create visual focal points with any plant that has a striking feature, like vibrant fall colour, stunning floral displays, colourful bark or berries
- express your style. Your design can be curvilinear or more geometric, emphasizing straight lines and shapes. They can be simple or complex. They can be open by choosing low-growing vegetation like lawns or perennial gardens, or intimate with large shrubs or a canopy of trees.
- create human scale, by creating a canopy overhead and breaking up large, open spaces
- complement other features, like defining an entryway to your home or pathway, using trees and shrubs or vines
- fit into their surroundings. Plantings on a large site viewed from far, should be grouped to form larger visual statements than on small sites viewed closer, where a single perennial can be distinguishable. The planting design should suit the scale, as well as architectural and natural features of its surroundings.
- capture the beauty of changing seasons. Choose plants that flower at different times to spread the bloom throughout the growing season. Leaf colour, texture and size can also create visual interest and contrast in the spring and summer, and can add vibrant warmth in the fall. Plants with colourful bark and berries can be chosen for fall and winter interest. Evergreen plants provide structure and interest all year, including the winter.

Figure 6-5: Evergreen shrubs and tall grasses provide colour and structure to this front yard during winter. Colourful bark, fruits and leaves, dried grasses and dried flowers that remain on certain plants all winter also provide winter-interest. Also, include structural elements like rocks.

b) Wildlife habitat

Plants, particularly native ones, attract a whole range of local wildlife, including birds, bees and butterflies, lured by their flowers, fruits, nuts, berries, seeds, and by the shelter they provide.

- When selecting plant species, diversity is the key to attracting wildlife. Select a range of species that fruit and flower at different times, and that provide food throughout the year. Include evergreens that provide protection from winter winds. Plants with different heights, from low ones to tree canopies, provide multi-layered shelter and food sources. Include both open, low areas and taller, denser areas. The transition zone between these areas is often a good place to view wildlife.
- Trees in particular provide multiple opportunities for refuge, perching and nesting as well as food sources from fruits, flowers, nuts, sap, wood fibre, leaves, etc. If your yard is too small for trees, plant vines.
- Look for the plant species and other features that work best at attracting wildlife in natural areas nearby.
- Some birds and insects are beneficial in dealing with pests. Refer to Chapter 7 for more information on beneficial species and reducing the impacts of pesticides.
- If you want to attract birds, avoid shearing plants into unnatural shapes that block their entry to the branches and shelter these plants provide.
- Fallen or dead plant materials, like logs and branches, also provide shelter and food.
- Water is an important element for attracting a host of species. For more information on wildlife in water gardens, refer to Chapter 4.

More details on woodland shade gardens or wildflower meadow or prairie gardens are provided later in Chapter 6 and 7.

c) Slope stabilization and erosion control

Plants help stabilize slopes and prevent erosion of the soil from wind and water running across slopes. Their root systems hold soil particles together, while the above-ground parts like leaves and stems slow down stormwater runoff and break the velocity of falling rain. Mulches also provide a cover to prevent wind and water from dislodging the soil particles. Plants that self-spread by rhizomes are the most effective. More details on slope stabilization are covered in Chapter 3. (Figure 6-7)

d) Places to play

Lawns provide a wear-resistant soft surface for walking and passive play, like tossing a ball or Frisbee. Trees and tall shrubs can also provide places to climb, hide or build a fort. Observing nature and learning about how plants grow is also an excellent educational opportunity for children.

Figure 6-6
Photo: Evergreen

Attracting butterflies

Select a sunny location, protected from the wind, that features water and certain flowering plants, preferably planted in large, easily visible groups. Here is a small sample: goldenrod, beebalm, Russian sage, New England aster, purple coneflower, butterfly weed, spirea, showy sedum, common yarrow, butterfly bush, lavender, black-eyed Susans, sunflower, cosmos and joe-pye weed.

Figure 6-7: Planting on a slope can not only create a richly layered effect, but stabilizes the slope and controls erosion, as described in Chapter 3.

e) Visual screens, privacy and access prevention

Plants, such as hedges, trees, large shrubs and vines can be used to create privacy by blocking views onto your property from outside. They are also used to conceal undesirable views, like an adjacent parking area.

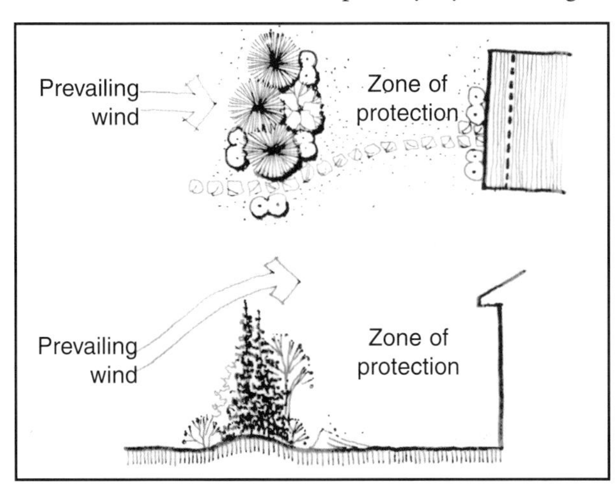

Figure 6-8: You can create a protected area behind a mixed screen planting that includes evergreen plants. Wind speed can be reduced by about 50 per cent up to a distance of about 15 times the height of the windbreak on the downward side. Most of the protection occurs within a distance of about 5 times the windbreak height. For example, for a 7.5 m (25 ft.) tall moderately dense vegetative windbreak, the maximum wind reduction occurs within 38 m (125 ft.) downwind of the windbreak or closer.

f) Block winds and save energy indoors

Plants are also used as windbreaks since they obstruct, direct, divert and filter the wind. They can diminish the strength of the prevailing winds and, when placed correctly, help conserve energy for heating your home. A mass of deciduous and evergreen trees and shrubs effectively reduces the velocity of the wind. Evergreen plants that branch to the ground are the most effective year-round windbreaks. A gap in the windbreak tends to concentrate wind flowing through the gap and create stronger winds by funnelling the air. When locating them, consider the direction of prevailing winds. (Figure 6-8)

Snow removal can be time consuming and costly. Rows of trees or shrubs can control where snow accumulates. Snow deposits behind windbreaks where wind velocities are decreased and is swept from areas with strong winds.

g) Provide shade and save energy indoors

On warm summer days, trees and large shrubs can provide shade and create the ideal place to quietly read a book, sip coffee or eat on the patio. By choosing and locating appropriate vegetation, you can save on your energy bills. The degree of energy savings depends on many factors including the tree's location, initial and ultimate size, growth rate, as well as sunlight transmission characteristics when in leaf and when bare.

If the space permits, plant deciduous trees to shade out summer sun inside your house and allow winter sun to penetrate through after the leaves drop. However, even a deciduous tree with no leaves can cast enough shade to block sun from a south-facing window in winter when you want the extra warmth. In winter, large deciduous trees can block over 50 per cent of the sun's rays, depending on the species. If reducing your home energy demand is a priority, trees located on the south side of the house should be a single-trunk, tall-growing species close to the house without lower branches that block winter sun (Figure 6-9 and 6-10). When the sun angle is high in the sky, like in summer, you may also not be shading the house at all if the tree on the south side is too low or too far away. Of course, the angle of the sun depends on your location, so determine the sun's angle at mid-day at various times of the year and locate your trees accordingly. For more details look for the CMHC publication *Tap the Sun*.

A more realistic location for shade trees is to the west and to a lesser extent, the east. Trees on these sides of the house can be shorter and further from the house than trees on the south side, making these more feasible locations for most homeowners. Locate the trees so they shade windows exposed to direct sunlight. If neighbouring buildings already shade them, the tree will have less impact.

Figure 6-9: The location of this tree, close to the southern wall of this house, provides shade on the roof and south wall in the mid-summer, helping to cool the house, while letting the sun penetrate through the windows in the winter when the angle of the sun is lower and the leaves have fallen.

A well-placed trellis with deciduous vines over a window can also shade out the sun indoors. If you have an air conditioner, shading it with plants will also help reduce your energy bills.

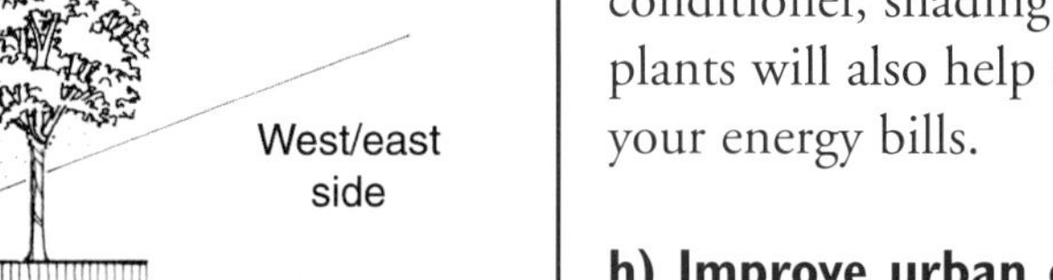

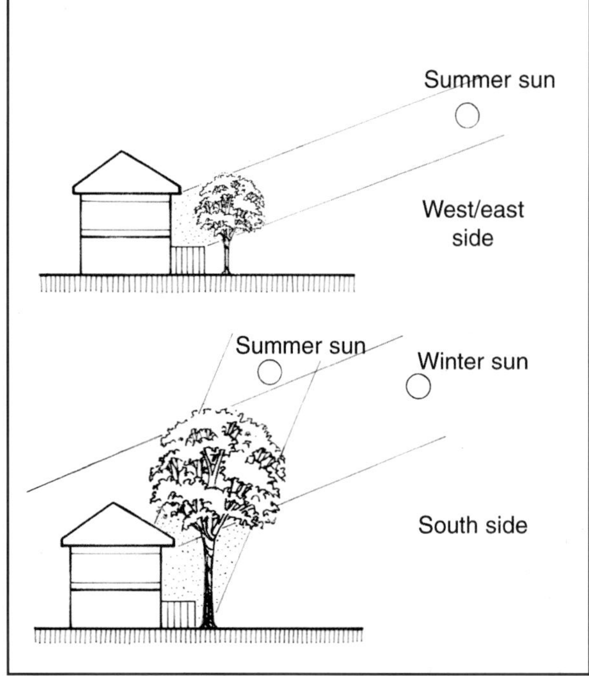

Figure 6-10: Deciduous trees block summer sun, and if properly placed to shade the house windows, walls and roof, help to keep the house cooler in summer but allow sun to warm the house in winter. Trees on the south side should be closer to the house and taller than trees on the west and east side, since the angle of the sun is higher on the south side.

h) Improve urban climate

In urban environments, vegetation has many important effects. It provides shade and it also dissipates water through evapotranspiration, thereby moderating or reducing ambient temperatures on hot summer days. By contrast, hard surfaces, such as parking areas and roofs, re-radiate solar radiation as heat and increase summer temperatures, at times by several degrees when compared to rural areas. Vegetation also helps to reduce wind velocities. These are good reasons to preserve trees and other vegetation in the urban environment.

3. Get to Know Your Site Conditions and Choose Suitable Plants

a) Plant hardy species that are suited to your region

Plants are ranked according to their hardiness in relation to Canada's large-scale climate variations. The "Plant Hardiness Zone" map (Figure 6-12) outlines the different zones in Canada where these plants will most likely survive, based on the average climatic conditions of each area, such as high and low temperatures, length of the frost-free period and summer rainfall. The production of the Canadian Plant Hardiness Zone map was a collaborative effort by scientists at Natural Resources Canada's Canadian Forest Service and National Atlas of Canada as well as Agriculture and Agri-food Canada. Refer to the Agriculture and Agri-food Canada Web site http://sis.agr.gc.ca/cansis/nsdb/climate/hardiness/ for more information or ask at a local garden centre for the most recent zone designation of your region and the rating of plants.

The lower the zone number is, the harsher the climate is. Each of these is then broken down into secondary zones "a" and "b". The "a" zones are slightly harsher than the "b" zones. The Arctic Zone is 0a, whereas the zone with the mildest climate in Canada (8a) is in southwest British Columbia.

Plants usually survive in zones that are the same or higher in number (more temperate) than those for which they were ranked. As an example, a red maple rated for zone 3 can be planted in a zone 3, 4 or 5 region without a problem. However, several factors also affect the survival of plants in a given zone, for example, exposure to wind, sun, heat, salt spray and air pollution. A backyard that is well protected from winds and that enjoys good snow coverage in winter can host plants normally living in a slightly higher ranked zone. However, its chances of survival are better if the plant is ranked for your zone or lower.

To choose plants that work in your region's hardiness zone, find out what zone your home is located in and refer to the plant identification

Figure 6-11

Hardy northern garden: Sutton-Fogwell residence, Yellowknife

Gerry Sutton and Lynn Fogwell's hardy northern garden adorns the yard of their Yellowknife house. Establishing the plants in the challenging climate required experimenting with shrubs and perennials that would tolerate the northern conditions, focusing on plants native to the Yellowknife region or elsewhere in the arctic as well as alpine species. They started most of the perennials from seed. Native plants include mooseberry, ferns, gooseberry, currants, potentilla, cloudberry, dwarf raspberries and alpine saxifrages. Delphiniums, bergenia, hardy rhododendrons, daphnes, prairie crocus, arctic poppies, several species of caragana and dwarf birches are also featured in the garden. Maintenance includes mowing the lawn and some pruning of shrubs. There is very little watering, the exception being brief and not common hot, dry spells. Sometimes perennials have to be moved or divided, but generally, maintenance is minimal.

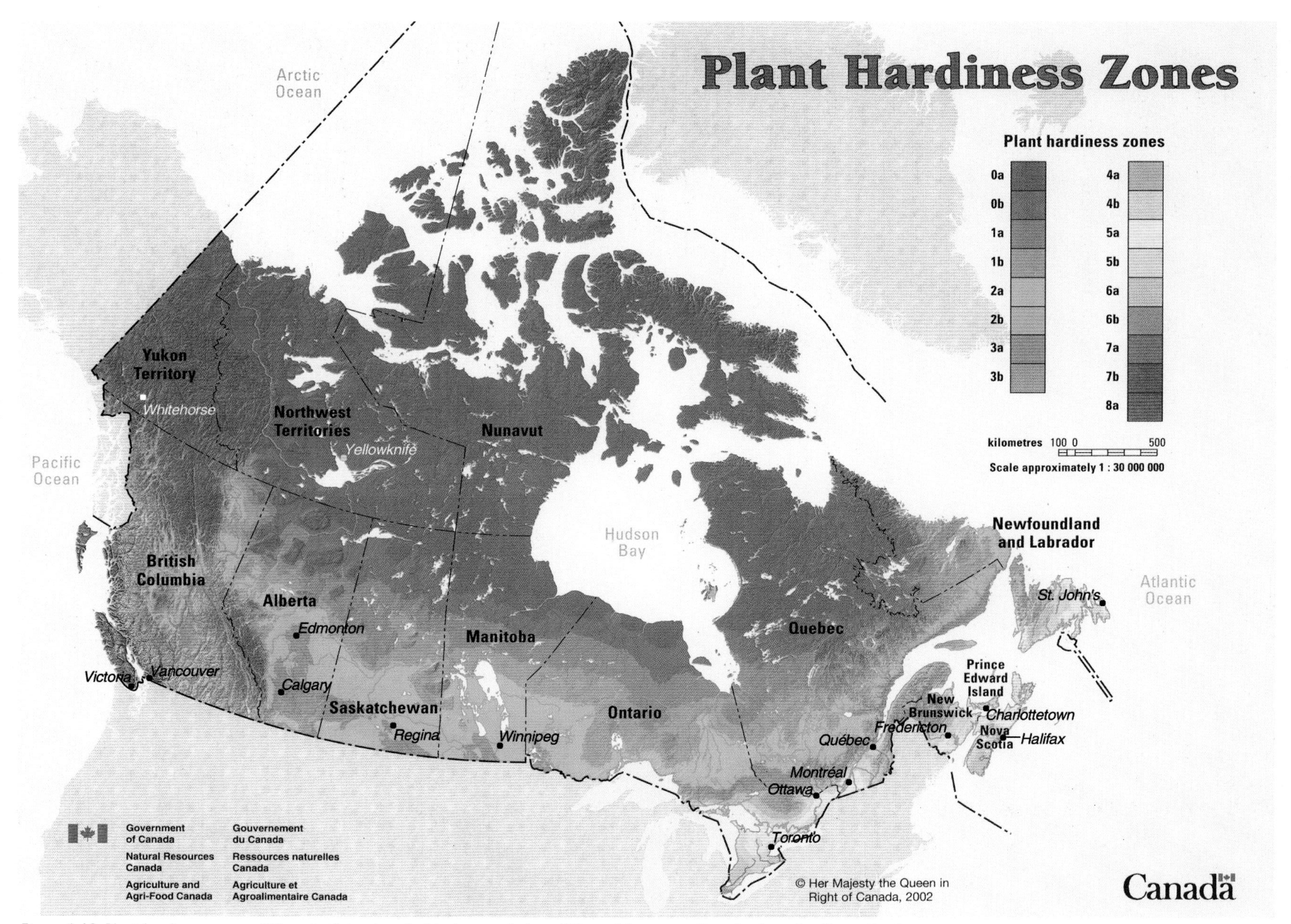

Figure 6-12: Plant hardiness zones of Canada. Reproduced with the permission of Natural Resources Canada, Canadian Forest Service.

Choosing species native to your region has several benefits: they are usually well suited to your local climate and planting them will help restore your region's native vegetation and its unique identity. Check the list of resources at the end of this guide for more information on native plants. But non native species should not necessarily be eliminated from your plant palette. If you select non-natives, focus on hardy species that are well suited to your site and non-invasive ones (see resource list for more information), especially in the vicinity of natural areas.

Generally, plants should not be dug up from the wild, unless you are rescuing them from a location where they would otherwise be destroyed, like a construction site, and you have permission from the landowner. If you're collecting seeds, obtain permission and take only 10 per cent of an individual plant's seed and only from species that are abundant. Ensure that they are accurately labelled.

labels at the garden centre or in plant catalogues. Otherwise, ask on-site personnel, as the hardiness zone is an essential element to know.

Plants with a higher hardiness rating may need to be protected in winter and may be more vulnerable to pests and diseases. This means more time and more money than planting species that will do well without pampering.

The source of the plants is also important. Indeed, a single species may sometimes cover a large territory. For example, snowberry, *Symphoricarpos albus*, grows from southern California to Alaska and from Quebec to Virginia. A plant that is hardy to your region may have been produced in a very distant place, much more to the south or been grown from a seed source from a different zone. It will therefore be more vulnerable to winter cold. Similarly, its needs for water may vary considerably. For example, a red maple from Tennessee will have a more limited chance of survival if planted in Ontario. Two individuals of the same species have a different genetic code. They have learned to live in and have adapted to the conditions in a given region. The closer they are planted to their point of origin, the greater their chances for surviving transplanting. Therefore give preference to locally produced plants that grow and live naturally in your area.

Learn to recognize plants

The use of the Latin names for the plants is the best way to identify them correctly. Many common names for a plant may exist depending on region and country. One common name may even refer to several varieties of plants. However, there is only one universally recognized Latin name for each plant.

Plants have a double Latin name. The first word refers to the genus and the second to the species. For example, in Latin, the sugar maple is called *Acer saccharum*, "Acer" meaning maple and "saccharum" identifying the species of maple. Furthermore, many species have been hybridized into several varieties or cultivars that will have different characteristics like height, shape and colour. An example is the "Dart's Gold" ninebark or *Physocarpus opulifolius* "Dart's Gold".

Planting plans and plant lists

Planting plans, like the one shown in Figure 2-4, identify the location, quantity and species of plants in your design. They are usually accompanied by a plant list that specifies important information about the plants, including:

- botanical name (Latin)
- common name
- total quantity of each plant
- size (such as diameter of the trunk, height)
- spacing (intervals between the plants)
- condition (for example, bare root, potted, balled and burlapped, wire basket)
- remarks, comments (planting, maintenance)

This is useful information to know when you are ordering plants and if you are installing them yourself. But a plant list is essential if you are hiring a contractor to do the ordering and installation. Below is a sample table you can use to start your plant list.

Botanical name	Common name	Quantity	Size	Condition	Spacing	Remarks
Acer saccharum	Sugar Maple	2	4 cm- (1½ in.) caliper	wire basket	3 m (10 ft.)	

b) Choose and locate plants according to your site conditions

Each plant has different needs for healthy development. This section describes the main needs.

Sunlight

All plants require light, but in different proportions. Many plants thrive in full sun and need it to produce flowers or fruit, while others burn if they are exposed to direct sun. They adjust better to shade or to the low-intensity light that you find in the shadow of buildings, fences and other plants.

Plants are usually divided into three categories: sun-loving plants, partial shade plants and shade plants. Sun-loving plants are those that require more than six hours of direct sunlight a day. The partial shade plants need to receive between three and six hours of sunlight to develop. The shade plants are generally content with less than three hours of filtered or direct sunlight. Not meeting the minimum amount of sunlight required by a plant could mean that the plant may not survive, may bloom less or appear to be unhealthy. Some shade plants will not survive if they are exposed to the sun; in other words, some species have shade tolerance, while others, including many woodland species, have shade requirements.

Soil

Soil provides the plants with the nutrients, oxygen and water they need to develop and to reproduce. The importance of preserving the existing soil and especially of planting vegetation that is suited to this environment, as well as amending the soil to suit the plant's needs was discussed in detail in Chapter 3.

Moisture

Each plant has specific water needs according to its physiological make-up or ecological niche. Plants are sometimes categorized according to three levels or degrees of moisture:

- dry, slight moisture
- intermediate, average moisture
- moist or wet

For plants to thrive, it is important that they receive the water they need. However, conserving water will save you money and time and reduce stress on municipal water systems and natural water sources, as discussed in several places in this publication. So to save water and meet the plant's needs, select plants that are suited to the moisture conditions in which you intend to place them. Don't forget that in the first full growing season or so of planting, they require more water than they do once they are established. It is critical for the long-term health to ensure they get adequate moisture during this time.

Different parts of your site will have different moisture conditions. High points and slopes will tend to be dryer than low points. North-facing slopes tend to be moister than slopes facing south or west. Areas exposed to wind and sun will also tend to be dryer than shaded and protected spots. Some soils can hold water better than others. It's important to know your site conditions before you choose your plants.

In addition to selecting plants that are suited to your local climate and specific site conditions, follow the design tips provided later in this chapter, the maintenance tips in Chapter 7 and the advice on irrigation in Chapter 4.

Space Constraints

Another important site consideration when selecting plants is the amount of space available. If you plant something that has a mature height or spread larger than the available space, you will end up needing to prune it, which requires more of your time or even costs if you need to hire someone.

c) Diversify and choose pest-resistant plants

Choose species that are resistant to pests, including insects and diseases. By choosing pest-resistant species, you can reduce the use of pesticides, which ultimately will save you time and money and minimize the impact on the environment. Also, diversify your choice of species—if there is an attack, only one species is affected. Keeping your plants healthy is also a good defence against attacks.

d) Other stress factors

For plants to thrive, it is important to minimize and consider their tolerance of certain stress factors, such as:

- exposure to wind and to frost in winter
- salts carried in the air or the soil
- air pollution
- vandalism
- snow removal and deposits
- soil erosion

Other factors may possibly jeopardize even the most vigorous plants. Construction and other wastes may contaminate the soil and affect drainage. Compaction caused by construction traffic, or storage of materials or by changes in level may also damage the root systems of existing vegetation, as discussed in Chapter 3.

4. Choose Low-Maintenance Planting Options That Work With Nature

We have described the functions that plants perform in the residential landscape. For example, trees and large shrubs provide shady, private environments protected from winds. Once the concept plan is finished, you will have located on the plan several clusters or large plants to meet specific functions. The choice of plants and their final positioning will be clarified as you draw closer to finalizing the landscape plan. At this point, you know roughly if and where you want to have trees, shrubs, lawn, low-growing groundcovers and floral displays.

Consider selecting a low-maintenance planting alternative that works with nature. In Chapter 1, we introduced the idea that working with nature allows you to save time, save costs and respect the environment.

What do we mean by a low-maintenance alternative? Although there are many exceptions and variations, the typical Canadian yard consists of a lawn, with a tree or two and a scattering of shrubs, perennials, bulbs and annuals. Maintaining these landscapes to high aesthetic standards can be a resource-intensive endeavour, in terms of time and costs and the following impacts:

- more water consumption. Municipal water consumption doubles in the summer, mostly because of lawn and garden watering. Many people over water their landscapes, water them when it is not needed or use inefficient watering techniques. This lowers water tables and reduces stream flows, which affects fish and other animals and also costs more for municipalities to supply and treat. It also increases your water bills.

Invasive species typically grow rapidly and have prolific seed production and dispersal systems to the point where they can aggressively spread into and overtake native species in nearby natural areas, such as woodlots, wetlands or ravines. Once established, they are difficult to control. Find out which plants are most invasive in your region or neighbouring ones; avoid planting them if you live near a natural area and if you wish to control their spread in your garden. For more information on invasive species, consult the resources at the back of this publication.

- increased use of electric and gasoline powered mowers, trimmers and leaf blowers that emit air pollutants and create noise.
- more use of pesticides. If you do have a pest problem, follow the tips and precautions provided in Chapter 7.
- increased use of fertilizers. Depending on conditions and usage, they can leach into ground water and penetrate into streams and lakes through stormwater runoff.

There is a growing trend toward more diversity and customization of plantings to individual homeowners' unique conditions and tastes. Many homeowners are looking to the planting alternatives described below.

Comparing different planting options

By recording the maintenance activities of 30 residential landscapes in southern Ontario over two growing seasons, CMHC compared the maintenance of different kinds of garden (refer to the table below). Each garden type has varied requirements in terms of time, cost, water, fuel, fertilizer and pesticides.

Each garden type is described in the sections that follow along with some results of the study. No doubt, maintenance depends on the individual homeowners' preferences: the amount of time you want to spend, and your aesthetic standards. Even among the garden types discussed below there will be huge variations in maintenance practices, but some garden

Relative comparison of landscape types for maintenance inputs

Landscape Type	Time	Costs*	Water-use**	Fertilizer use	Pesticide use***	Energy use	Benefits
woodland shade garden	low	low	med.	none	negligible	none	- excellent for wildlife habitat, biodiversity, stormwater infiltration, shade, privacy, erosion control, improving urban climate - natural look
wildflower meadow	low	low	low	none	negligible	low	- excellent for wildlife habitat, biodiversity and colourful displays - good for stormwater infiltration, erosion control, improving urban climate - natural look
low-main-tenance lawn	lowest	low	none	low	none	med.	- stormwater infiltration, erosion control, improvement of urban climate - wear-resistance surface for activities
xeriscape	med.	med.	low	med.	low	none	- good for stormwater infiltration, erosion control, improving urban climate - attractive, colourful displays - more traditional look
ornamental flowerbed	high	high	high	low	med.	none	- good for stormwater infiltration, erosion control, improving urban climate - attractive, colourful displays - a more traditional look
conventional lawn	low	med.	med.	high	high	high	- stormwater infiltration, erosion control, improvement of urban climate - wear-resistance surface for activities

* Includes plant replacements/additions, mulch and other material costs as well as hiring maintenance companies.
** Includes watering of new plants as well as established plants.
*** Includes organic and synthetic pesticide use.
The data shows relative maintenance inputs from a study of 30 residential landscapes in southern Ontario. The landscape types are described below providing more details on the study results. Adapted from *Residential Landscapes: Comparison of Maintenance Costs, Time and Resources,* CMHC, 2000.

types are naturally predisposed to requiring less maintenance.

Low-Maintenance Lawns:

Lawns are the dominant landscape feature of many Canadian homes. They withstand foot traffic and are inexpensive to install. However, they can be a high-maintenance endeavour, but don't have to be. Two ways you can reduce the time, costs and environmental impact of a lawn is to limit its size to what you need for recreational activities and to opt for a low-maintenance lawn as described below. When we recorded the maintenance requirements of lawns, we looked at two types:

- Conventional lawns: made up of a small number of fine turfgrass species like Kentucky blue grass. To keep them in a green, homogenous, manicured state, many people neatly mow them at least weekly and regularly water, edge, fertilize and treat them for pests (disease, insects, weeds). Again, these tasks can vary greatly from one lawn to the next, depending on the site and your preferences.
- Low-maintenance lawns: composed of broad selections of hardy, drought-tolerant, slow-growing turfgrass and wear-tolerant broadleaf species such as white clover that do not require frequent mowing, watering or fertilizing. They look less uniform or smooth than conventional lawns, but generally function in the same way.

For a comparison of turfgrass species, refer to Table 7-3 in the next chapter.

In the maintenance study described earlier, the low-maintenance lawns took 50 per cent less time, 85 per cent less costs, 50 per cent less fuel, 100 per cent less water, 100 per cent less pesticides and 85 per cent less

Figure 6-13: Close-up comparison of conventional lawn (left) and low-maintenance lawn (right)

To cut back on pesticides, you can include a diverse mix of species in your lawn. If a less homogenous look is acceptable to you, consider a low-maintenance lawn that has a broader range of species so it is more resistant to pests. Using a mix of turfgrass species that is suited to your site, as well as some wear-tolerant broadleaf species, can also help reduce watering, mowing and fertilizing needs.

Figure 6-14: Low-maintenance lawn: Other than mowing it a couple of times a year, the owners of this summer home do nothing to maintain their lawn. It contains white clover, moss and turfgrass species. Over the years, it has withstood many games of croquet and mini-putt. Photo: Cathy Bastien

In Chapter 7, you will find some low-maintenance tips for lawns. For example, often lawns are cut closer and watered more than they need to be. Set your mower height to cut the lawn no lower than 6 to 8 cm (2 ½ to 3 in.) in height, for most grasses. This will help shade the roots and make them better able to hold water. This saves you time and gasoline or electricity. Established lawns normally require about 2.5 (1 in.) of water per week to thrive. If Mother Nature is providing this amount in rainfall, there is usually no need for supplemental watering. If you live in an area that experiences periods of low rainfall or drought, you may want to consider a low-maintenance lawn that would include low-water use species. Or you may want to assess whether you really need a lawn and opt for a low-water use alternative like a xeriscape or wildflower meadow (described below).

Figure 6-15: Shade-tolerant groundcovers are a low-maintenance option in shady places, like under trees or in narrow side yards. The density of this well-established bed helps keep out weeds.

fertilizer per year to maintain than the conventional lawns. Installation costs would be roughly the same, although low-maintenance seed mixes or sod may be a bit more expensive and harder to find than conventional lawn seed mixes or sod.

Other lawn alternatives:

In some heavily shaded areas, like a narrow side yard, slopes or low spots where foot traffic is not anticipated, you may want a low-growing planting that is not labour intensive. A wide variety of shade-tolerant groundcovers, are available at nurseries. These can be low-maintenance alternatives to lawn.

Try moss, particularly for shady, damp areas. Attempts to make grass grow in this type of environment are often fruitless. Look in Chapter 7 for details on how to grow moss.

Figure 6-16: This small 4.6 m^2 (50 sq. ft.) urban backyard in Toronto is home to a diverse range of woodland species. Plants in this multi-storey woodland shade garden are well suited to their environment and mutually supportive. For example, the ferns and Solomon's seal thrive in the shady understorey of the alternate-leaved dogwood and serviceberry and help keep the soil rich, moist and well drained by returning organic matter as they decompose. Photo: Jim Hodgins

Woodland Shade Gardens:

Like lawns, the species of trees and shrubs you use and how you arrange them can affect your maintenance time and costs as well as impact on the environment. The maintenance study looked at two approaches:

- Ornamental trees and shrubs: featuring species that are selected primarily for their floral and foliage displays and form, such as crabapples and hybrid roses. Depending on the species, location and individual preference, many homeowners regularly prune, water, fertilize and treat them for pests and rake their leaves. They are often surrounded by lawn and separated from other trees and shrubs as isolated species.
- Woodland shade gardens (Figure 6-16): composed of hardy native trees, shrubs and ground covers, like ferns, that mimic natural forests as follows:
  - plants are grouped together, protecting each other from exposure to the drying effect of wind and sun
  - the soil is covered by living and decaying plant materials such as mulch and leaf litter. This improves soil moisture and aeration, reduces weed growth and erosion, and adds nutrients to the soil. The cycle of

Woodland shade and pond garden: Skelton residence, Vancouver

Photo: Frank Skelton

Figure 6-17

Frank and Erin Skelton have a delightful woodland and pond habitat garden at their home in Vancouver. Because it features species that are native to this region of British Columbia and adapted to the climate and site conditions, very little maintenance is required. Some plants have been added and moved over the 20 years since its creation, but much of the garden has evolved naturally without human intervention, as the trees grow and light, moisture and soil conditions change. Maintenance is generally limited to cutting off the fern fronds in spring, some weeding and pruning plants that encroach on the pathway. The only pest problem has been slugs that were eliminated with ferric sulphate, a non-toxic fertilizer containing iron. Frank and Erin only water new or recently relocated plants until they are established. Rainfall takes care of the rest.

The water in their pond is recirculated with a pump that is cleaned and oiled twice a year; the pond is cleaned once in the spring to clear out the leaves. This small effort is well worthwhile, since the sound and sight of falling water creates a remarkably tranquil environment. The birds and other creatures like dragonflies and water striders attracted to this habitat also add to the tranquility and help to keep the mosquitoes down. Plant species include: deer fern (*Blechnum spicant*), dagger-leaved rush (*Juncus ensifolius*), bunchberry (*Cornus Canadensis*), queen's cup (*Clintonia uniflora*), wild ginger western species (*Asarum candatum*), Labrador tea (*Ledum groenlandicum*), salal (*Gaultheria shallon*), snowberry (*Symphoricarpos albus*), red huckleberry (*Vaccinium parvifolium*) and western laurel (*Kalmia microphylla*).

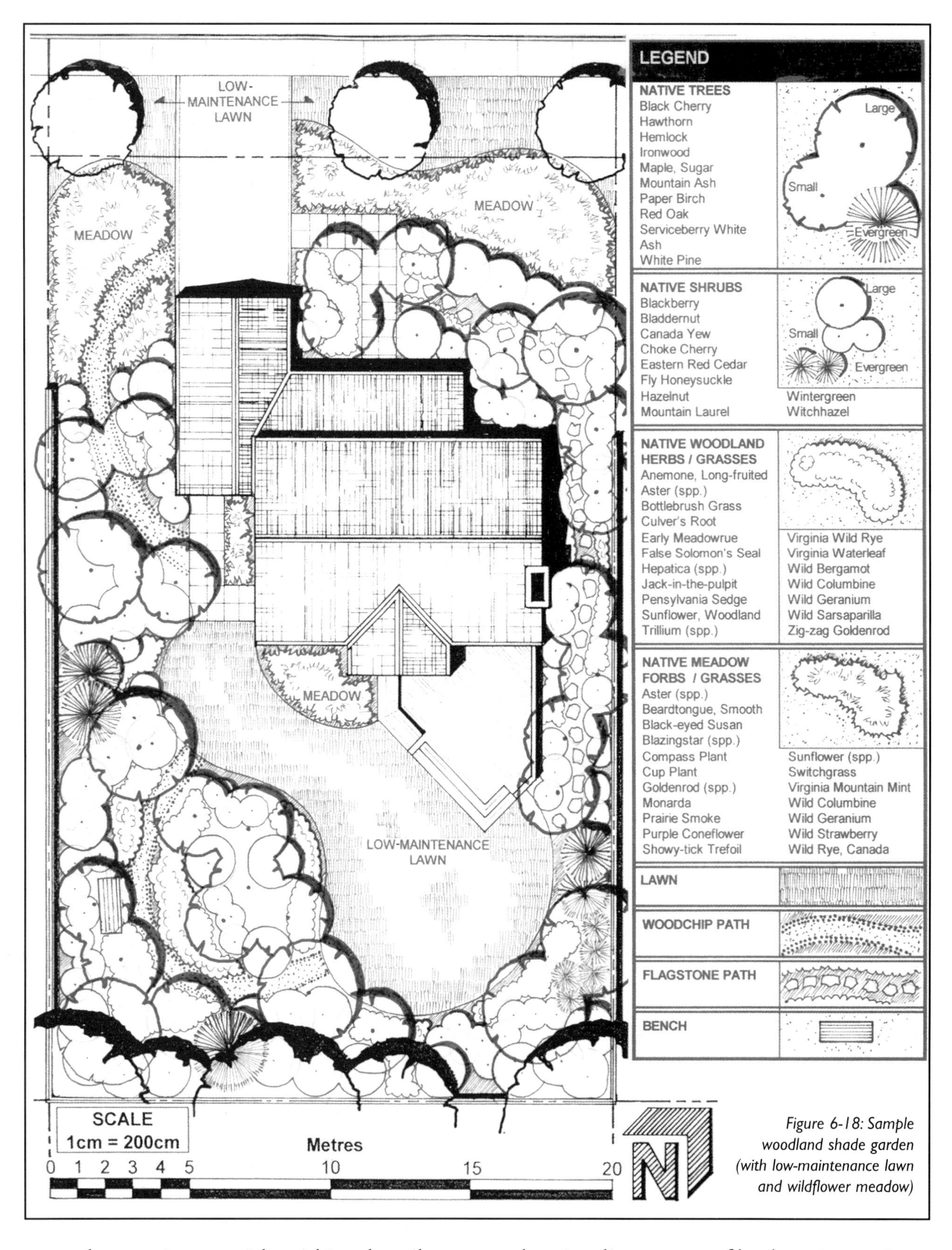

Figure 6-18: Sample woodland shade garden (with low-maintenance lawn and wildflower meadow)

decomposing material enriching the soil and life forms in the soil continues as new leaves and plants grow and die. All of this reduces the need for watering, fertilizing and weeding.

- there is a diverse range of hardy native species, adapted to their environment, from the upper tree canopy providing shade to the shrubs, seedlings and ground cover plants below. Diversity allows the planting to better withstand pests. It's a mutually beneficial plant community that relies more on natural processes to thrive, rather than human input, but actual time spent in the garden depends on your preferences.

- they provide a shady, informal, natural look and attract birds to your backyard. The natural look means that you prune less.

Installation and maintenance tips are provided in Chapter 7.

In the maintenance study, woodland shade gardens took about 75 per cent less time, 60 per cent less water, 100 per cent less pesticides and fertilizer and cost 80 per cent less to maintain per year than the gardens featuring ornamental trees and shrubs that were studied. Installation costs were roughly the same. Figure 6-18 shows a sample design featuring a woodland shade garden.

Figure 6-19: Wildflower meadow. Photo: Randy Penner

Wildflower Meadows and Tall-grass Prairies:

The species of perennials, ornamental grasses, bulbs and annuals you select and how you arrange them can affect your maintenance time and costs as well as impact on the environment. The maintenance study looked at two approaches:

Working with Natural Processes: Forest Succession

When left to natural forces, landscapes evolve and change. In Canada's forest regions, disturbed sites, like an agricultural field, are colonized by seeds that tolerate exposed conditions, and the site evolves into a grassy meadow. As soil becomes more fertile and less exposed, seeds from new pioneer species emerge, like poplars that grow rapidly and are content with poor soils. Soil, light and moisture conditions then become more favourable to species like maple and beech that seek a more shady, moist, fertile environment. The plants naturally wait for their desired growing conditions before establishing themselves. Eventually, the site evolves into a forest with a tall canopy of trees and understorey of plants in the rich soil and shady environment below. The mature plant community will differ from region to region in Canada. In some regions it will be a mixed forest of conifers and hardwoods. These plant communities are adapted to local conditions, like soil, rainfall, sunlight exposure and temperature, and as a result are self-sustaining. Even in an urban backyard, a small wildflower or woodland shade garden can thrive with little maintenance.

In nature, succession from meadow to old-growth forest can take roughly 100 to 300 or more years. When establishing a woodland shade garden on your property, you can significantly speed up the initial stages of the process by planting woody species normally found in more mature stages of succession and nurturing them along in the first few years with water, mulch and weeding. Depending on the species, plant size and layout, an attractive closed-canopy woodland shade garden could be achieved in about 4-15 years.

- Ornamental flowerbeds are the dazzling borders, beds and rock gardens of colour featuring perennials, bulbs and annuals that many homeowners regularly weed, water, deadhead and so on. They often include bulbs and annuals that require replanting every year. Again, these tasks can vary depending on the gardener and the site.

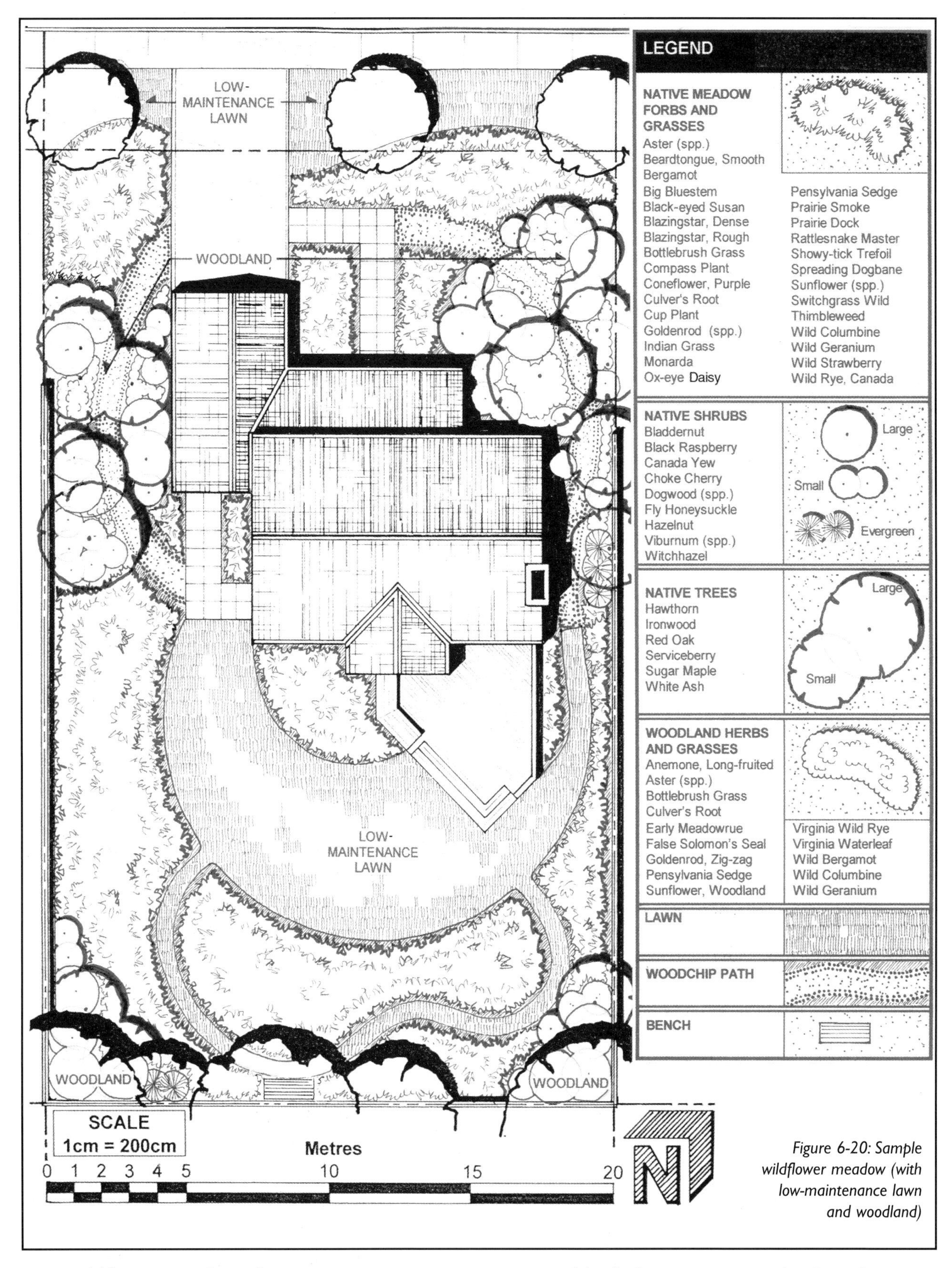

Figure 6-20: Sample wildflower meadow (with low-maintenance lawn and woodland)

- Wildflower meadows feature grasses and wildflowers that are adapted to their environment and mimic natural meadow or prairie landscapes, offering a subtle, natural look that attracts songbirds and butterflies. This is not an unmown lawn gone wild. You seed and/or plant seedlings selecting species that are suited to your

Wildflower garden: Penner Residence, Winnipeg

Janet and Randy Penner have been enjoying their backyard prairie garden at their home in Winnipeg for over 10 years. They established the garden by planting plugs of a wide range of native Manitoba wildflower and grass species in a 50/50 ratio: half grasses, half wildflowers. There are over 40 different species. Grasses include green needle grass, side oats grama, little bluestem, prairie dropseed, blue grama, Canada wild rye and big bluestem. Wildflowers include wild bergamot, purple prairie-clover, wild crocus, blanket flower, golden alexanders, smooth aster, meadow blazing star, giant hyssop, culver's root, closed gentian, hedysarum, cutleaf anemone and indigo bush. Plants were selected based on height, light and moisture preferences, time of bloom and color. They were planted in a pseudo-random fashion so that all the members of a given species would be scattered around rather than appear as one solid block of plants.

Photo: Randy Penner *Figure 6-21*

Rock paths running throughout the garden provide interesting play opportunities for the children. The garden never looks the same from month to month or year to year. It is at times predominantly yellow, while at other times, purple blooms predominate and transform the look of the garden. With fall comes a palette of rust, brown and gold. Aroma is another important aspect of the garden. After midsummer during humid weather, the yard is redolent with a pot-pourri of scents from such species as the wild bergamot and prairie dropseed.

Little maintenance is required. They mow the garden once every year in spring, collect the clippings and either compost them or burn them on a friend's property in the country (burning is illegal in the city) and spread the ashes on the garden. They mow in the spring rather than the fall because they enjoy the look of the tall grasses in the winter. Some spot weeding is required to get rid of any thistles and dandelions that show up. Otherwise, this lovely, dynamic garden maintains itself.

site and preferably native to your region. They work best in full sun. Maintenance can be as little as mowing once per year. Installation and maintenance tips are provided in Chapter 7.

In the maintenance study, wildflower meadows took about 85 per cent less time, 90 per cent less water, 100 per cent less fertilizer, 95 per cent fewer pesticides and cost 95 per cent less to maintain per year than ornamental flower beds we studied. Some gasoline or electricity was needed for the once annual mowing of the wildflower meadows. Installation costs were roughly the same, but could be greatly reduced in the wildflower meadow by seeding all or part of the area instead of planting. Figure 6-20 shows a sample design featuring a wildflower meadow.

Xeriscapes (Water-Saving Landscapes)

Xeriscapes are water saving alternatives that can include trees, shrubs and herbaceous plants, like perennials and ornamental grasses. These water-efficient landscapes look similar to conventional gardens but are made up of species that are suited to local rainfall conditions and that

Figure 6-22: This xeriscape features low water demand shrubs and perennials, like thyme, lavender and sage, as well as a mulch of pea gravel. Photo: Patricia Chinell

require little watering. Plants are grouped in mulched beds according to their water needs.

In the maintenance study, annual maintenance for the xeriscape design shown in Figure 6-23 was compared with another design for the same home that featured a conventional lawn, ornamental flowerbeds and ornamental trees and shrubs. The xeriscape took about 25 per cent less time, 90 per cent less water, fertilizer and pesticides and cost 60 per cent less to maintain per year than the more conventional design. Installation costs were roughly the same.

You can create a lush, colourful, low-maintenance, water-saving garden by applying the following principles of xeriscapes:

1. Get to know your property

First, get to know the moisture conditions on your property where you want to plant by conducting an inventory and analysis plan of your property, as detailed in Chapter 2 and the soil analysis tips in Chapter 3. The physical limitations of your property, such as slopes and drainage, will also affect water needs. For example, place plants that prefer moist conditions in low areas rather than on slopes where you will have to water them more often.

Also consider shade and drying wind. Shade reduces evaporation and therefore the need to irrigate. You can reduce water losses in lawns by almost 50 per cent by moving from full sun to shade.

A shaded area can be as much as 10°C cooler than an area in full sun, and as a result, there is less water loss from the area. Some plants thrive in hot, dry areas, while others cannot tolerate full sun or dry conditions. Match your plants with your site conditions.

2. Use plants that suit the local conditions including rainfall, soil type and drainage

Choosing the right plant for the particular location is the key. Consult your local garden centre for plants that are suited to rainfall conditions in your region. Plants that are adapted to local climate and your specific site conditions can better withstand water shortages. The plant list at the end of this guide indicates plants that are well suited to a xeriscape. Conifers are generally more resistant to dryness than deciduous plants. Established plants have lower water requirements than recently installed ones.

3. Group plants according to their water requirement

Water can be used more efficiently by identifying "hydrozones." Rather than uniformly watering all your plants, cluster them into different zones that will be irrigated differently, according to the plants' water needs and site conditions. In the majority of cases, the most irrigated zones are those that are most visible, such as the main entrance to your home. The moderate water-use zones contain plants that require some watering during hot, dry periods. Low-water use zones require little or no additional water after the plant is established.

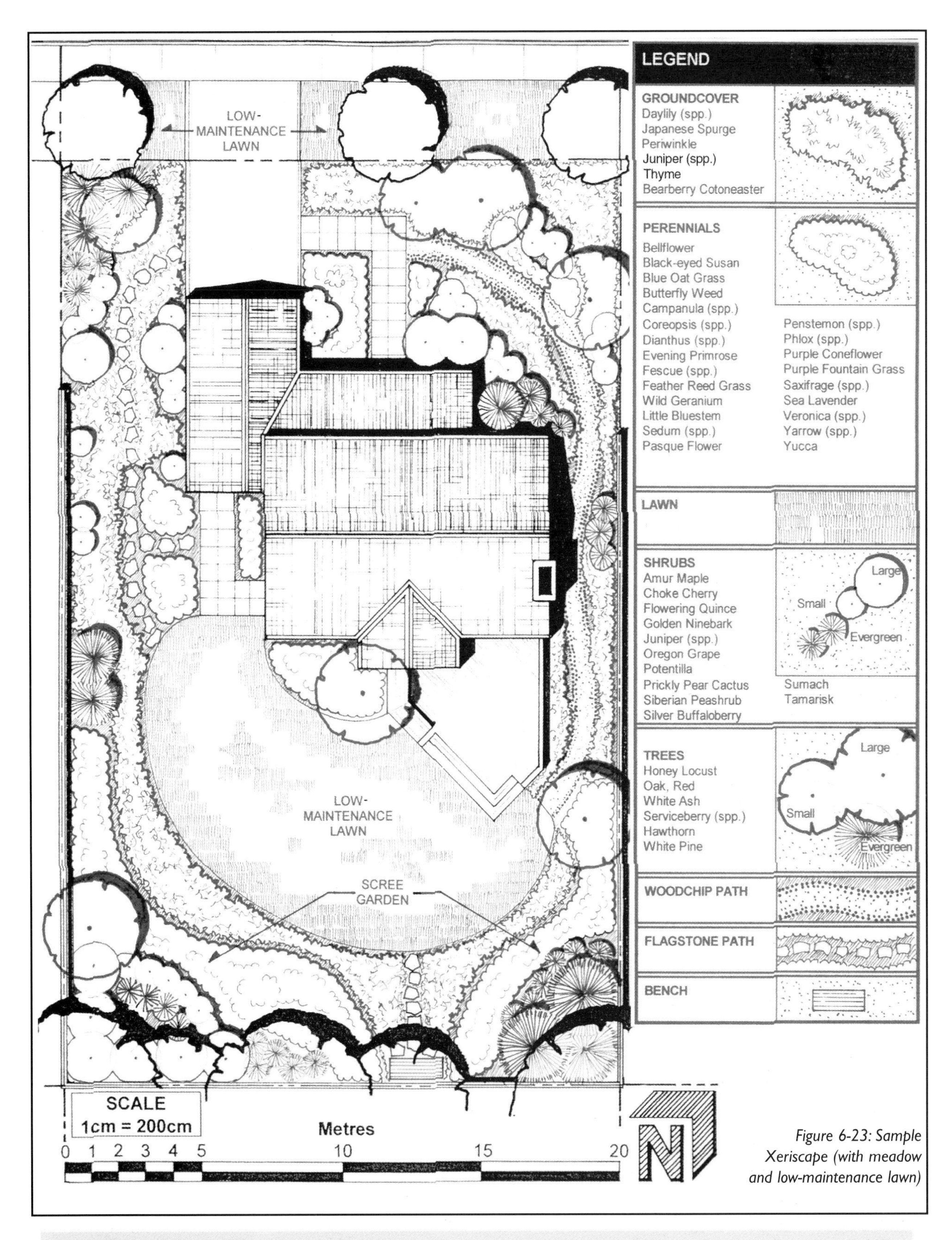

Figure 6-23: Sample Xeriscape (with meadow and low-maintenance lawn)

Consolidate your lawn area and size it for your social and play needs: Avoid many small or narrow lawn areas in favour of one or two consolidated areas. This way, watering, mowing and other maintenance activities can be done more efficiently. For footpaths or narrow spots, such as side yards, consider alternatives to lawn, such as wood chips, decorative pebbles, crushed stone or natural stone. These options require no water at all.

4. Improve the water retention capacity of the soil

Chapter 3 provides tips on how to analyze your soil and amend it accordingly. One of the most important improvements is the addition of organic material. It helps sandy soil hold nutrients and water, and improves aeration and drainage in clay soil. It also moderates soil temperature, which helps reduce water loss. All this results in better moisture conditions in the soil and healthier plants that cope better during hot, dry weather. One of the most economical sources of organic matter is compost. You can improve your soil by adding organic matter every year as well as during planting.

5. Use mulch

Mulch is a must for any water-efficient landscape. It discourages weed growth, moderates the temperature of the soil, decreases evaporation and improves soil moisture. It also protects the soil from runoff and wind, thereby reducing erosion. Organic mulch decomposes slowly into the soil, improving soil aeration and can provide nutrients to the plants. Refer to Chapter 7 for a comparison of mulch types and other mulch tips.

6. Water efficiently

Many homeowners waste water by using inefficient irrigation practices, over-watering and missing the opportunity to capture stormwater. Chapter 4 describes different irrigation systems and compares them for efficiency. The most efficient irrigation methods apply water directly to the soil and roots, such as soaker hoses or drip irrigation systems. These are used in planting beds, vegetable gardens and under trees, shrubs and hedges. For lawns, well-designed in-ground sprinkler systems apply water efficiently. Timers and moisture sensors also increase water efficiency. Follow the watering tips in Chapter 7, such as knowing your plants' water needs and avoiding over-watering. When plants are over-watered, roots cannot absorb the surplus water, which runs off towards the sewer and is wasted. Also, water during the coolest periods of the day and above all not in full sun, since a large percentage of the water will be lost to evaporation.

Vegetable Gardens/Food Production

Many people have productive food gardens in their front, side or back yards or on decks, balconies, or patios. Food producing plants do not have to be confined to a garden plot and can serve other functions. For example grape vines on a trellis above your deck can provide shade on warm summer days, privacy and visual appeal. So can a pear, apple or walnut tree. Many attractive shrubs, such as currants or serviceberries, that might appear in an ornamental bed or in a foundation planting, do bear edible fruit, which is commonly used to make jam. Herbs like thyme, sage, rosemary, basil or oregano can be very attractive additions to a flowerbed or border.

Plant list

The plant list at the end of this guide is only a small sample of plants for the Canadian garden. Consult the resources listed at the back of this book for more information.

Xeriscape: Mason Hogue residence, Uxbridge, Ontario

Photo: Marjorie Mason Hogue Figure 6-24

During the winter of 1993, Marjorie Mason Hogue moved into a new home in Uxbridge, Ontario. The following spring, much to her chagrin, she discovered that her two-acre property was an immense sand lot. Worse still, the existing well supplied only a small amount of water.

She made a decision; a large portion of her land would become a xeriscape. The large south-facing, sandy slope would be planted with pines, junipers, dwarf spirea, cinquefoils, daphnes and other plants tolerant of dry conditions. Perennials such as stonecrops, thyme, lavender and ornamental grasses completed the picture, adding touches of colour. Only small plants were planted. They were situated based on their specific needs (soil, watering, sunlight, etc.) As a result, after eight years, more than 95 per cent of the plants were thriving, without any watering or spreading of fertilizer!

Xeriscape: Barone Oehmichen residence, Oka, Quebec

Photo: Friedrich Oehmichen Figure 6-25

In 1984, Sandra Barone and Friedrich Oehmichen purchased their home in Oka, Quebec, a beautiful country house on an immense piece of land. Owning only a manual lawnmower, they quickly eliminated more than 50 per cent of their front lawn.

For a maximal biodiversity, the area was divided in two zones. In the first, they brought in sand and created a large dune where they put in a garden of plants that are adapted to dry and poor soil conditions. Now, lavender, Russian sage, yarrow and blue oat grass gently wave in the wind.

In the second zone, they created a garden with plants that attract hummingbirds. The existing soil was preserved and amended with home compost, and minimal-maintenance, drought-resistant plants were given preference. In effect, almost no watering has been necessary during all these years.

PLANTING METHODS AND MAINTENANCE

Chapter 7

An important part of having a cost- and time-efficient landscape that minimizes environmental impact is to determine how best to install and maintain it. This chapter provides construction details and maintenance tips for trees, lawns, shrub and flower beds, hedges, woodland shade gardens, wildflower meadows and vegetable gardens.

1. General Tips for Reducing Your Garden's Maintenance Needs and Environmental Impact

No matter what type of garden you design, you can save time and money dedicated to maintenance by being inspired by nature and responding to the conditions on your site. The choice of plants as well as the planting method affect how much maintenance will be required. Here are some tips that can be used to reduce maintenance:

- Probably the most important advice is to select plants that are suited to the places you want to put them. For example, find out the mature height, spread and shape of the plants to ensure that they fit the space. If they are too big, you will need to prune and that can be time consuming. For tight spaces, select the compact variety of a shrub or tree. Ensure that the water needs of the plants you select are suited to the moisture conditions of the specific site—is it on a dry slope or a wet low point, is it in a sunny or shady location, what type of soil do you have, how much rainfall do you get? Find out the sun and soil requirements of the plant and match them to your site conditions. (Figure 7-1)
- Use only hardy plants that are well suited to your local climate. (Figure 7-2)
- Use flowering shrubs and perennials to create floral displays instead of annuals and bulbs that need to be replanted every year. If you use annuals, choose self-seeding species. If you use bulbs, use perennial bulbs that do not need to be dug every year and try naturalizing bulbs that multiply on their own.
- Use mulch. It helps to reduce weeding and watering, moderates soil temperature and reduces erosion. Organic mulch also improves soil aeration, moisture and fertility.

Figure 7-1: For sunny locations select sun-loving plants, like the ones in the top photo: garden phlox, blanket flower, perennial geranium and black-eyed Susans. For shady locations, for example, under trees, select ferns or other shade-tolerant species.

- Select a diversity of disease resistant species.
- Consolidate planted areas according to their water and other needs so that maintenance can be done efficiently, particularly for lawns. For example, watering and mowing one consolidated area of lawn is more efficient than several small patches scattered throughout your property.
- Define edges of planting beds with borders to help contain the mulch and soil and reduce maintenance, if you want a clearly-defined look. Refer to Chapter 5 for a comparison of border materials.
- Adopt the general maintenance tips described below, like the advice for watering and integrated pest management.

Figure 7-2: Hardy trees and shrubs that are well adapted to the soil and climate conditions, like jack pine, trembling aspen and white spruce, surround this home in Yellowknife.

2. General Maintenance Tips

Watering

Often water is applied inefficiently and not taken up by the plant due to over-watering, evaporation or runoff. Here are some general watering tips to help avoid wasting water:

- Before watering, always take into account the amount of water supplied by rainfall to your lawn or garden in the preceding week and the water requirements of your plants. Leave a measuring container in your yard to help monitor rainfall, but empty it once per week. Feel the soil for moisture and look for signs of drying, like wilting.
- Bear in mind any watering restrictions that may apply in your municipality.
- Water in the early morning before 9 a.m. to reduce evaporation and scorching of plant leaves.
- Water on calm days to prevent wind drift and evaporation.
- Set up your sprinkler or hose in such a way that it avoids watering hard surfaces such as driveways and patios.
- Apply water directly to the soil, not to the leaves. It's the roots that absorb water, so aim for the ground.
- Ensure that water absorbs deeply into the soil, rather than applying superficial but frequent sprinklings. Thorough and deep watering encourages deeper roots.
- Water slowly to avoid runoff and to ensure the soil absorbs the water.
- Regularly check your hose or irrigation equipment for leaks or blockages.
- Collect rainwater from your roof in a rain barrel, and use a bucket or hose to apply it to your garden. Keep the rain barrel covered with an insect screen. Chapter 4 also describes water conservation techniques.
- Choose an efficient irrigation system. Drip irrigation systems are highly efficient because they deliver water slowly and directly to the soil. This promotes deeper roots, which improve a plant's drought resiliency. A soaker hose placed on the ground at the base of planting beds, vegetable gardens and under trees, hedges and shrubs applies water to the soil surface where it is needed and reduces evaporation. Well-designed in-ground sprinklers are an efficient system

Table 7-1: Some Non-synthetic Pest Controls

Name	Damage	Some control methods
aphids	suck sap from leaves, twigs, stems or roots. Leaves develop yellow spots and may wilt or curl.	Spray with soapy water (10-25 ml non-detergent soap in 4 litres of water). Spray plant undersides weekly with jets of cold water. Spray fruit trees with dormant oil in spring, while dormant. Place pans of water with yellow food colouring nearby to attract and drown them. Empty them once per week. Attract ladybugs and lacewings to your garden by placing wood stakes coated with a sugar liquid around the garden.
forest tent caterpillars	feed on deciduous trees	Destroy the egg-cases and cocoons by scraping them off with a knife. Cut off infested branches and burn or crush individual webs. Wrap burlap around the tree at chest height, tied in the centre, then folded in half. Check daily for caterpillar larvae that crawl under it.
chinch bugs	cause yellow patches on grass, which can turn brown and die	Aerate the lawn in spring to reduce compaction, remove thick layers of thatch, water the lawn thoroughly in dry periods and keep it over 6 cm tall. Big-eyed bug and tiny wasp are natural enemies of chinch bugs. Put 30 ml of dishwashing soap in 7 litres of water and drench a small area of the lawn. Chinch bugs will crawl to the surface to escape. Lay a flannel sheet on the treated area, wait 10-15 minutes. They will crawl on the sheet and get trapped. Vaccum them or drown them in a bucket of water.
voles (field mice) and mice	eat bark from young trees and shrubs	Wrap the base of trunks of young trees with a plastic spiral, a small mesh or piece of drainage pipe with one end in the ground.
earwigs	can damage leaves, fruits, vegetables and flowers. They frequent dark, moist areas.	Place pieces of garden hose or rolled newspaper in the vicinity of the earwigs. Place an inverted flowerpot filled with moistened straw or newspaper. Partly fill a tin can with fish oil or vegetable oil and bury it to the rim. Shake the traps described above into hot water.
slugs and snails	leave large holes and sticky deposits mainly on leaves and herbs of garden vegetables	In the evening, use a spoon to dislodge them then place them in a container of soapy water or rubbing alcohol. Lay down an inverted grapefruit, melon peel or flowerpot. Place boards covered with aluminum foil along plant rows, foil side up. During the day, they will take shelter there. Check the traps daily and place slugs and snails in a solution of soapy water or rubbing alcohol. Place some shallow containers (e.g. aluminum plate) buried to the rim, and partially fill them with beer. They will crawl in, bloat and die. Change the beer once or twice a week, especially after a rain. Sprinkle sand, wood ashes or baked eggshells at the base of plants to deter them from climbing the plants.
white grubs	feed on roots of grass causing lawns to wilt and turn brown	Clean up the lawn and garden in fall. Remove excess thatch and aerate compacted areas in lawns. Mowing height should be 6-8 cm. Leave lawn clippings. If grubs are detected in warm dry periods, water and fertilize. Applying nematodes dilluted in water in late summer may also be effective. Larkspur and geraniums may be toxic to the grubs. Choose resistant varieties, e.g. some fescues and ryegrasses.
gypsy moth	eats tree leaves	Scrape off egg masses into a bucket of hot water and household bleach or ammonia. Hand pick and crush larvae and pupae. Wear gloves when handling the insects. For caterpillars, wrap a 45 cm wide strip of burlap on the trunk, tie a string around the centre of the burlap and fold the upper portion down. The larvae will hide in the fold. Later in the day, lift the burlap, remove the larvae and dispose as described above.
white pine weevil	attacks at least 20 tree species, including eastern white pine	Prune infested trees in late July or as soon as possible after signs of infestation. Prune close to the top most unaffected whorl of branches. Burn or destroy pruned branches and leader stem. Banding tree trunks and the base of leader stems with sticky tape may stop adult weevils from reaching the leader stem.

Adapted from Pest Notes, Pest Management Regulatory Agency, Web site http://www.hc-sc.gc.ca/pmra-arla/english/consum/pnotes-e.html

for lawns. Timers and moisture sensors also increase water efficiency. Chapter 4 includes a comparison of irrigation systems.

- Do not over-water. It can weaken the plant's root system and carry nutrients down into the soil and away from the plant's roots.
- Water back from the tops of slopes, as water will run down the slope and seep into the soil.
- Select plants that are well suited to local rainfall and the water conditions of your site.
- Keep new plantings moist until they establish (usually for the first and, in some cases, second growing season).

Pests

Many Canadian municipalities educate their citizens on reducing pesticide use and many are considering or already have restrictions or bans on the cosmetic use of pesticides. These initiatives have arisen out of concerns about the potential health and environmental risks of pesticide use.

Pests can include insects, weeds, diseases and fungus. The best way to deal with pest problems is to prevent them. But even with good gardening practices, pest problems can arise. **Integrated Pest Management** is an approach that uses a combination of strategies, putting prevention at the top of the list:

- **Prevention:** Choose the right plant for the conditions on your property and select plants that are resistant or tolerant of the pest problems common in your area. Your local garden centre or plant catalogues can help with this. Keeping plants healthy will help them resist attack and compete better with weeds. Keep weeds down by using mulch and by avoiding bare patches of soil that weeds can germinate in. Having a diverse mix of species will also help reduce the impact if one species is attacked. Refer to the many tips in this and the previous chapter on selecting, designing and maintaining healthy plantings.
- **Monitoring and proper identification:** Keep an eye out for pests or pest damage and make sure you correctly identify the pests and what is causing the problem. Learn about them, what they like or dislike and when they are most likely to be encountered. Consult your local garden centre or the resources at the back of this book for help.
- **Decide what you can live with:** The mere presence of a pest does not necessarily mean it's a problem. Treat it only when your monitoring shows that the pest is building up to an unacceptable level.
- **Select the least harmful treatment:** Treat problems as they arise, not automatically or on a regular basis. Select the least harmful, most specific treatment for the pest you are trying to control. This could include manual or cultural controls, for example, removing affected plants and plant parts when you see them. Pulling weeds by hand is another example, as is trapping of certain pests. Plant certain species that repel pests or attract the natural enemies of pests. Biological controls use natural enemies of the pests.

It's important to know the insects and animals, and their role in your garden. Some are beneficial, so be careful not to eliminate them. Beneficial ones include: bees, ladybugs, small wasps, praying mantis, spiders, beetles, green lacewings, bats and birds.

Only consider a pesticide product after following the above advice, including prevention, monitoring, proper identification and use of non-synthetic control methods, like the ones described above. If these approaches fail, and you feel you can't tolerate the pest problem, follow these tips when using a pesticide:

- Treat only specific problems as they arise, not automatically or on a regular basis.
- Check with your municipality to see if there are any regulations restricting or banning the use of pesticides.
- Use the pesticide only on the pest it is intended to control, in the exact location of the problem.
- Spot treat where the problem exists only rather than treating the entire lawn or garden.
- When in doubt, contact an informed professional with knowledge of ecological or organic landscape techniques.
- Follow instructions and safety precautions on the label. Compare labels and warning symbols of different products.
- Be careful not to kill useful critters like ladybugs, dragon flies and earthworms.
- Wear protective clothing and wash your hands thoroughly after handling any pesticide product.
- Never spray in windy conditions.
- People and pets should not come into contact with the treated surfaces until residue has dried completely. Tell your neighbours ahead of time.
- Don't dispose of pesticides down the drain or storm sewer. Call your municipality for hazardous waste disposal sites.

Soil Amendments

The best thing is to select plants that are suited to your soil conditions. But if some adjustments are necessary, find out what kind of soil is needed for your plants and look in Chapter 3 for tips on soil analysis and amendment. Adding organic matter to the soil will improve its water holding capacity in sandy soils, drainage in clay soils, aeration and fertility. A low-cost source is compost, grass clippings and fallen leaves (more tips on composting later in this chapter). Maintaining a layer of organic mulch over your plantings also improves the soil.

Power Equipment

Most maintenance equipment is available in different versions: manual, electric or gasoline-powered. Among all the models, gasoline engines pollute the most. If you own one, check it regularly and adjust it so that it operates as efficiently as possible. For small lawns, a manual mower can be used to reduce air emissions and noise as well as getting some exercise to stay healthy. The same can be said for manual hedge trimmers and rakes for collecting leaves. Mulching or composting mowers finely shred grass clippings and leaves so they can provide a natural source of nutrients without matting or restricting air and water movement.

There may be some equipment that you use very infrequently. A rototiller, for example, might be used a couple of hours a year. To reduce your own costs, resources used to produce the equipment, and waste when discarded, consider sharing such equipment with neighbours.

Mulch

In planting beds and underneath trees, mulch is indispensable. It controls weeds, moderates soil temperature protecting roots from the cold and heat, reduces erosion, decreases evaporation, and improves soil moisture. These advantages save you money and time, particularly for watering and dealing with weeds.

Mulch is either organic, derived from living material, like bark, straw or wood chips, or inorganic, like pebbles. Organic mulch decomposes into the soil, improving soil structure (less compacted, more workable, better aeration), improving moisture holding capacity of sandy soil and drainage of clay soils, and mulches like compost, increase soil fertility. Since organic mulches decompose, they need to be topped-up to a sufficient depth regularly. Many of the organic mulches, such as wood chips, are by-products or recycling of other materials. This helps to reduce landfill waste.

Table 7-2: Comparison of Mulch Materials

Material	Description	Note-worthy features
ORGANIC MULCHES		
1. compost	- produced by composting leaves, grass clippings, bark and kitchen waste (e.g. fruits and vegetables)	- excellent for soil structure, moisture and nutrients - adds organic matter quickly - if available at home, saves transportation energy and costs - free, if you compost at home - decomposes quickly—needs to be replenished most frequently
2. bark chips/chunks (e.g. pine, cedar)	- are generally the by-products of milled wood - see Figure 7-3	- decorative - slowly decomposes - suppresses weeds well - good resistance to wind and compaction
3. wood chips, (e.g. cedar, hemlock and pine)	- generally made from tree trimmings, Christmas trees, wood waste - see Figure 7-22	- colour may grey over time - less expensive than bark chunks or shells - suppresses weeds well
4. hulls or shells (e.g. cocoa)	- made from shredded shells or hulls that are the by-products of various crops	- decorative, fine texture - suppresses weeds well - lightweight—prone to blowing in strong winds
5. straw	- can be bought from local farmers - straw mulch blankets held by degradable mesh are also available	- suppresses weeds well - light-weight—easily carried by the wind if not held together by mesh - low cost
6. shredded leaves	- collect from your yard or ask a neighbour - best if finely shredded and allowed to sit for several months	- if available at home, saves transportation energy and costs - free if you have trees at home - leaves need to be shredded finely, for example, with a mower, or they can mat together and block air and water movement
7. grass clippings	- collect from your lawn or ask a neighbour - allow to dry before applying - best if finely chopped	- if available at home, saves transportation energy and costs - free, if you have a lawn - adds nitrogen to soil - using a mulching mower helps chop them finely to prevent matting and blocking air and water movement
8. pine needles	- collect from pines on your property or ask a neighbour	- if available at home, saves transportation energy and costs - free, if you have pine trees - acidifies soil - slowly decomposes
9. newspaper	- lay it down in layers and anchor with soil or stones or shred it. Use black and white paper only.	- least attractive—use under more attractive mulches - available at home, free
INORGANIC MULCHES		
1. landscape fabric (geotextile, weed barrier)	- single layer of fabric available at garden centres - cut slits to plant through	- used under other mulches like wood chips, pebbles or stones
2. decorative pebbles, stones, lava rock and other loose aggregate materials	- available at garden centres - see Figure 6-22	- decorative - longest lasting—does not decompose - good wind resistance - not for use in vegetable garden - use over landscape fabric to reduce mixing with soil

Before spreading the mulch, remove existing weeds. Mulch depths can vary depending on the mulch type, but generally apply about a 5-7.5 cm (2-3 in.) layer of mulch. Layers of fine-textured mulches like cocoa shells or compost can be thinner than fluffier mulches like straw and shredded newspaper. Too much mulch may be harmful to the plants due to reduced oxygen reaching the roots. Also be careful to keep it away from the base of plants at trunks and stems to prevent diseases such as crown and stem rot.

Having mulch under trees rather than lawn will help eliminate competition for water and oxygen and prevents damage to the tree bark that can be caused by lawn mowers and trimmers. Ensure mulch does not contact the trunk. Mulch that is mounded up against the trunk can lead to bark rot, which can cause disease and insect problems.

Figure 7-3: Mulch helps reduce weeding and watering, moderates soil temperature and reduces erosion. Organic mulch, like this bark mulch, also adds organic matter to soil, improving soil fertility, moisture and structure.
Photo: JMK Imag-ination

In areas where you are encouraging relatively self-sustaining ecosystems, like woodland shade gardens, natural debris, like fallen leaves, will eventually take the place of added mulches.

Although mulch has numerous benefits, there are a few cautions. Watch for toxic ingredients and for sour mulch—mulch that has not been properly composted. Although uncommon, it can damage or kill plants. It smells like vinegar, ammonia or sulphur and is very acidic. Good mulch smells like fresh-cut wood or garden soil. In a particularly wet season, mulch can become a host for molds, some of which can be harmful to plants. Lift mulch to check for molds and remove mulch if needed.

Composting

Composting is a natural process that transforms organic matter into a product that looks like soil, called compost. To compost, you need air, humidity and micro-organisms such as bacteria.

Composting your green waste is easier than it might seem. You can help improve the environment by keeping almost a third of your waste from going into landfill. In addition, the compost produced from your green waste will be an excellent soil amendment to be used in your garden and plant beds. It's also free. Some municipalities offer domestic composters at very low cost and also produce their own compost, available to the public at low cost.

Most organic matter can be composted at home. It is essential that you use an appropriate mix of organic matter to obtain maximal results. The following table presents a summary of waste to use or avoid.

Gardening Waste	Kitchen Waste	Do Not Use
• leaves (they decompose faster when shredded) • plants and soft plant stems (not weeds with seeds) • pine needles • bark, wood chips, sawdust • old potting soil • grass clippings, preferably dry	• fruit waste • vegetable waste • tea bags • coffee grindings and filters • bread, toast • shredded paper • crushed egg shells	• meat, fish and bones • dairy products • sauces • grease and oil • animal excrement • cold charcoal or ashes

To successfully compost your green waste, follow the advice of your municipality and the following points. You will obtain rich compost after only 10 to 12 weeks.

- Look for a sunny, well-drained part of the yard that is easily accessed from the kitchen year-round. Turn over the soil in the location you have chosen for the composter. Contain your heap with a covered composter, for example, a wood frame box with wire mesh or a bottomless, plastic composter with a lid and retrieval door. After installing the composter, cover the bottom with a layer of branches to improve drainage and air circulation. Start with a layer of moist material, such as kitchen waste, followed by a layer of dry material, like leaves or yard waste. Then add a layer of soil to start decomposition.
- Composting is more effective when the pieces of organic material are small. Plant stems and kitchen waste should be cut up. Finely chop leaves and grass clippings.
- Do not add a thick layer of a single type of waste. Do not put more than 6 cm (2½ in.) of grass clippings or 15 cm (6 in.) of leaves in a single layer.
- Let grass dry before putting it in the composter, or mix it with dry material, such as leaves, to prevent it from compacting.
- Turn the pile every 5-10 days or when you add material to allow air to circulate.
- Use a cover that locks in place to keep nuisance animals out.
- The contents of your composter should be slightly moist, like a squeezed-out sponge. If it starts to smell, it may be too wet. Turn the pile. If the material is too dry, it will take a long time to compost. Water it.
- Dig in food waste each time you add it or cover it with soil to keep away odours, insects and animals.
- When adding yard waste, ensure that weeds do not have seeds. They could germinate in your garden.

It is also possible to compost in the winter. You can put green waste in the composter all winter, but the decomposition process will be slower or will stop when the pile is frozen. The composting process will begin again in the spring and a good turning will reactivate the process.

Figure 7-4: Composting reduces household waste and is a cost-free, renewable way to improve your soil.

Here are some solutions to the most common problems:

- When the pile does not diminish or produce heat, you will need to stimulate the process. If the pile is dry, add water and mix it well. If the pile is damp and muddy, spread it in the sun and add dry waste. Keep compost to remix with raw materials.
- When the centre of the pile is damp and hot, but the overall pile remains cold, enlarge the pile. Keep the composter as full as possible. Mix old waste with new waste, and dry waste with damp waste, breaking up any clumps.
- When the pile is damp, has a sweet smell, but does not heat, it probably lacks nitrogen. If this is the case, add cut grass, kitchen waste and a bit of organic fertilizer.
- If the compost has a bad odour, give it air by breaking up any clumps to loosen the pile, unplug the vents and add dry materials like wood chips. Turn the pile to promote aeration.

The compost is ready to use when it is dark, crumbles or gives off an earthy odour. You can screen the compost to eliminate matter that has yet to decompose. Put this matter back in the composter to continue the process.

Use your compost in the vegetable garden and other plantings. It will help to increase soil fertility, structure, aeration and porosity, making it easier to work with and improve its ability to retain water in sandy soils and improve drainage in clay soils. This will help you to grow healthy plants.

TREES

Planting Methods

Choose trees and their location according to their needs, like soil, sunlight and moisture, as well as space constraints, like overhead wires and underground structures. Consider the tree's mature height and size. If it's for a tight spot, select a narrow canopy tree so that you won't have to prune as much. It is important to mention that tree roots spread beyond the outer tips of the tree's crown, so be careful to leave enough space for the root system to properly develop. Consult plant catalogues, ask at the garden centre or follow the directions on the tags for soil preference, height, spread, hardiness zone and fruit, flower and leaf information.

As with all plants, choose tree species that are well suited to your soil type. Wherever possible, use the existing soil to fill the planting hole. Analyze your soil and determine what amendments are needed following the information in Chapter 3.

Spring and fall are the best times to plant balled and burlapped, wire basket or potted trees. Bare root stock should only be planted while the tree is dormant, before leaf-out in the spring and for certain species, after the leaves drop in the fall.

Mark the tree's future location. Dig a hole that is at least twice the width of the root ball. The hole should be as deep as the root ball, so the top of the soil in the pot or the top of the root ball is at the same level as the surrounding

Trees can be bought from the garden centre in different forms:

- containers
- wire baskets or balled and burlapped (enables larger sizes, harder to lift) or
- bare root (lighter and easier to lift, cheaper, plant only when dormant, for very young trees, e.g. seedlings, and shrubs. You can inspect the roots. Roots must be kept moist before planting). See figure 7-6.

At the garden centre, look for:

- plump, healthy buds
- healthy undamaged bark and branches free of abrasions, insects etc.
- leaves that are not yellow or otherwise discoloured (see note below for bare root)
- well-spaced, firmly-attached branches
- weed-free soil around the roots

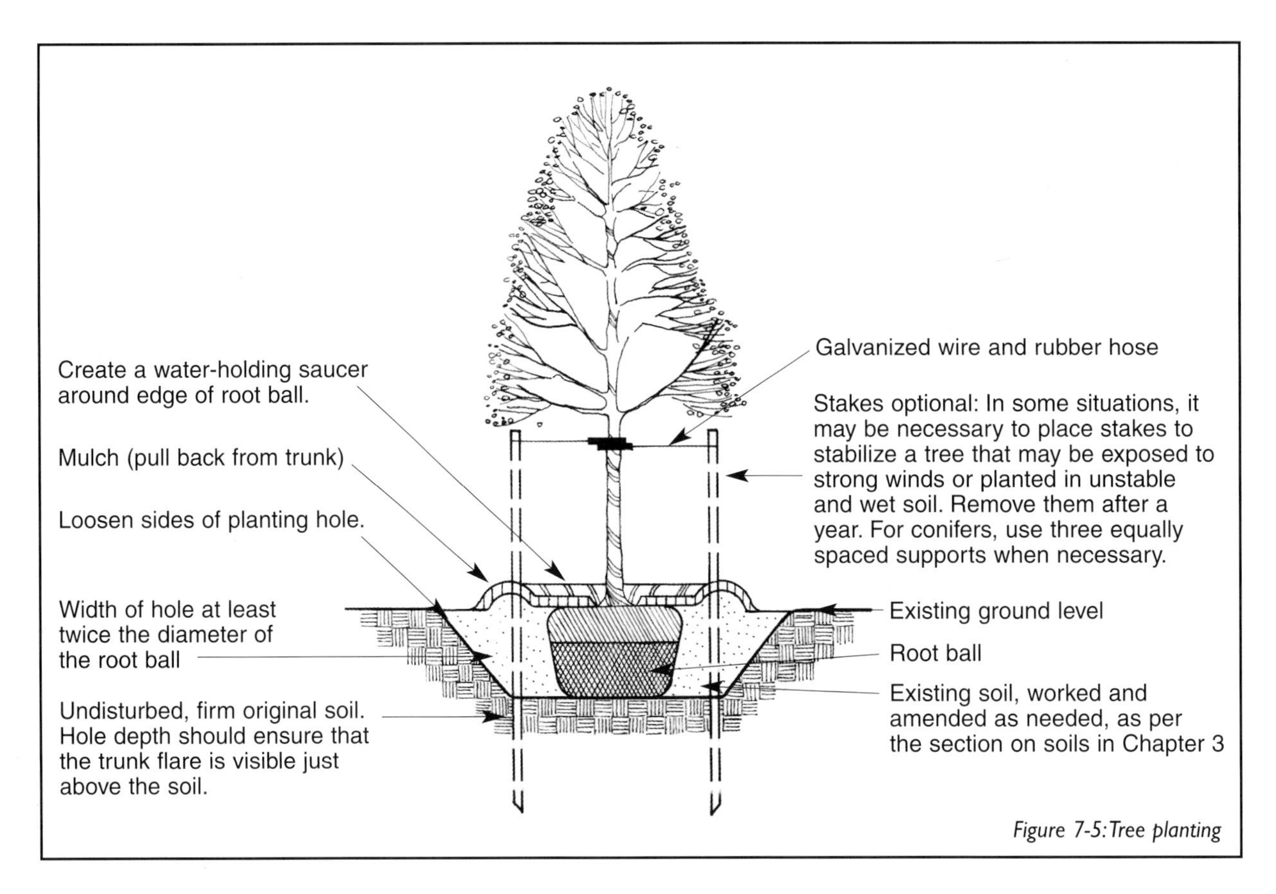

Figure 7-5: Tree planting

soil. The trunk flare (the swelling where the roots connect to the trunk) should be visible at or just above the finished level of the soil when the root ball is placed in the bottom of the hole. Look under the burlap to see if the trunk flare is visible. The bottom of the hole should be firm and preferably unworked to help avoid soil settlement, which can cause the tree to be too deep in the soil, which can lead to problems. Loosen the soil on the sides of the planting hole.

Water the root ball before planting. For potted trees, remove the pot, and place the tree in the planting hole. For balled and burlapped or wire basket, put the tree in the hole, then cut the rope or wire without damaging the root ball. You can simply pull the fabric back from the top 1/3 of the root ball and leave it to decompose in the hole. Leave the fabric and rope or wire under the root ball in place to avoid damaging the root ball. Stand back and view the tree from different angles to ensure that it is straight.

Fill the hole with 20 cm (8 in.) layers of soil and tamp it lightly with your foot as you place the soil. Once filled, form a water-holding saucer to direct water to the roots. The inside of the saucer edge should be the width of the root ball. Water abundantly but slowly so that the water penetrates deep into the soil. Fill the saucer with water once and then a second time. Spread a 5-7.5 cm (2-3 in.) layer of mulch over the entire surface of the saucer and pull it back from the trunk so it does not come into contact with the trunk. Refer to the section on mulch earlier in this chapter.

Keeping the soil moist for the first or second growing season is critical. One rule of thumb is to water trees with a one-hour trickle at least once a week, barring good rainfall, and more frequently during hot weather. Continue until mid-fall, tapering off for lower temperatures that require less frequent watering. During the second year, water twice a month during late spring and summer. Check soil moisture periodically, particularly during a drought and water abundantly (slowly and deeply) if it is dry. Adding a root stimulant, like mycorrhizae or bone meal, encourages rooting and healthy plant growth. Prune dead branches.

Transporting and storing plants

Transporting plants should be done in a closed vehicle to prevent drying. They should also be protected to prevent damage when working with them. Plants should be delivered just before you are ready to plant them. If they are stored temporarily, keep the roots moist and protect the plants from the drying effect of the sun and wind. For transplanting and storing bare root stock, refer to the text box on bare root plants.

Staking or other means of stabilizing trees is not usually necessary. In most cases, unstaked plants will form a better quality root network to keep the tree in place. However, if the soil is unstable or wet, or if there are high winds, plants should be stabilized. Stakes or turnbuckles are removed as soon as the tree is well rooted, usually a year after planting.

Tree Maintenance

Fertilizing and pests

If your tree is well suited to your soil, you should not need to apply fertilizer or amend soil on an ongoing basis. Organic mulch, including fallen leaves, will help keep plants healthy, as will applying compost in the spring before replenishing the mulch. Consult your local garden centre for advice if you notice growth problems.

For pests, refer to the general information on integrated pest management earlier in this chapter, including prevention, monitoring and proper identification, and using the least harmful treatment. Weeds will compete with newly planted trees for nutrients and water and may shade out small trees or seedlings: so keep weeds around new trees under control. Although mulch or

Bare root plants

Planting bare root plants should be done when they are in a state of dormancy. For deciduous trees, this means early spring, before leaf-out or autumn after the leaves fall. For conifers, you should wait until the fall. Take precautions when transporting bare root stock. Roots should be kept constantly moist and exposed as little as possible to the sun and wind until the minute you are ready to plant. Moistened fabric or shredded paper around the roots or digging a shallow hole and covering the roots with sawdust or moist soil can help prevent problems.

Before planting, trim damaged branches. Dig the hole as described above and place a small cone shaped mound at the centre. (See Figure 7-17 for bare root hedge planting.) Place the plant, gently spreading the roots over the mound, ensuring that the trunk flare is visible at or just above the finished level of the soil outside the hole. Fill the hole with soil ensuring that the trunk flare is at or just above the soil level. Create a watering saucer 10 cm (4 in.) high around the edge of the hole. Follow watering directions as described above. Larger bare root trees should be stabilized with stakes for the first year. Apply a 5-7.5 cm (2-3 in.) layer of mulch over the soil, being careful to keep it from touching the trunk. Refer to the section earlier in this chapter on mulch. Rodent guards may be needed to protect the young bark and trunk against rodent damage.

Figure 7-6: Planting a bare root tree. Photo: Evergreen

groundcover plants will help reduce weeds, you may need to remove them for the first few years, for example by hand pulling them. Once a shady canopy is formed, weed growth will be reduced.

Pruning

If you have chosen healthy trees with mature height and size that fits their location, pruning should be minimal. This will save you time and money in the long run.

Before pruning, consider these objectives. First, remove material that is dead, damaged, dangerous or diseased. After dealing with these problems, you can prune established trees for other reasons, such as to correct structural deficiencies, for example crossing or rubbing branches, or insufficient light and air in the centre. More light increases flower and fruit production.

Removing leaves reduces photosynthesis and may reduce overall growth. This is why pruning should be done sparingly. You should not remove more than 25 per cent of the crown. Most routine pruning, like removing diseased or dead branches, can be done any time of the year. Timing depends on the species, but generally, the best time is before the spring growth flush. Avoid heavy pruning just after the spring growth flush as this can stress the tree.

Cutting a straight clean edge at the branch collar is critical because broken branches and loose or torn bark can harbour insects and disease organisms. Torn and damaged bark should be removed. Use a sharp tool. Avoid cutting too close or flush to the trunk, which will open it up to infection and slower closure. Do not cut into the branch bark ridges (rings or lines of bulging bark that may be rougher and darker than surrounding bark), since this zone is an effective barrier to decay between the branch and trunk. If you wish to shorten a branch, cut just above a bud or secondary branch.

Branches over 2.5 cm (1 in.) in diameter should be cut with a saw. Remove them by making three cuts to avoid splitting and bark tearing:

1) Cut on the bottom side of the branch 30 to 60 cm (1-2 ft.) from the branch attachment, one-quarter of the way through.
2) On the top side of the branch 2.5 cm (1 in.) out from the first cut (away from the branch attachment), saw until the limb falls off.
3) Saw just beyond the outer portion of the branch collar. This can be done in two phases, initially from the bottom halfway. through and then from the top.

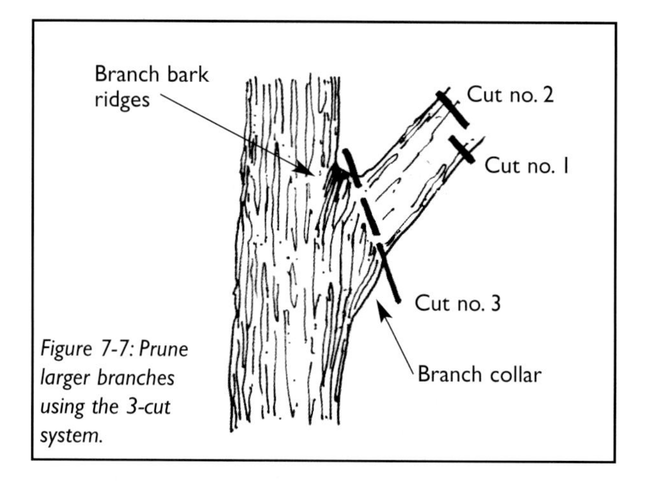

Figure 7-7: Prune larger branches using the 3-cut system.

Clippings from diseased plants should be burned to reduce the risk of spread. To avoid the spread of disease, disinfect the blades of tools using 70 per cent alcohol.

For some tree problems, it's best to hire a certified arborist, particularly if branches are hard to reach or if pruning involves working above the ground or using power equipment. This will also help you make informed decisions about your trees.

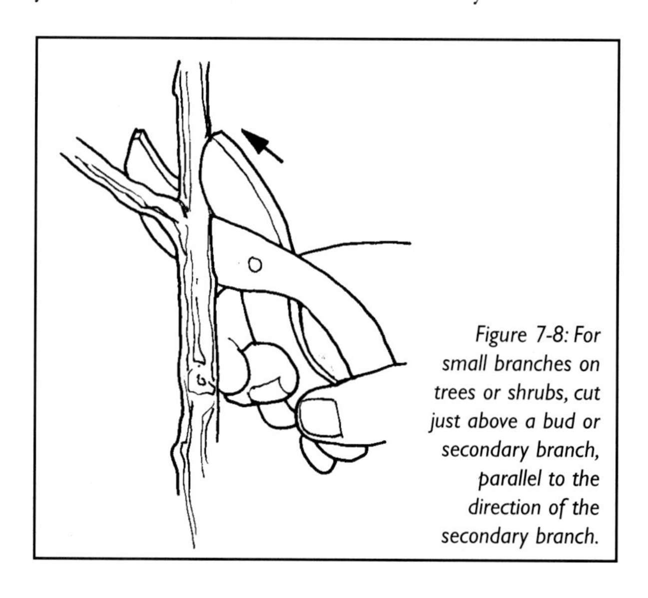

Figure 7-8: For small branches on trees or shrubs, cut just above a bud or secondary branch, parallel to the direction of the secondary branch.

For more information on pruning, refer to CMHC's *About Your House* CE12 on *Helping Your Trees Survive Storm Damage* and to the resources section in this book.

Watering

Plants have different water needs according to their species and stage of growth. For the first or second growing season after planting trees, they should be abundantly watered so that the soil is kept moist (see above). During this time, regularly check the moisture levels in the soil your trees were planted in, especially during drought periods, to supply them with the water they need. Once the plants are established, if the species is well suited to the site conditions, they usually require little or no water other than that provided by rainfall. But water during periods of low rainfall or drought or the long-term health of the tree may be affected:

- Apply water where it will reach the roots that take up the most water. These are generally located within the top 30 cm (1 ft.) of the soil. They are near and beyond the drip line (the area directly below the outer tips of the branches). Apply water around the drip line.
- An easy way to water trees is to place drip irrigation or a soaker hose (see Chapter 4) around the drip line of the tree and water until the area is saturated to a depth of 20-25 cm (8-10 in.).
- Water deeply to reach the feeder roots. A superficial watering encourages superficial root growth, which can make the tree more sensitive to drying-out and disease.
- Follow the general watering tips listed at the beginning of this chapter.

Raking leaves

Save time raking and bagging leaves by letting them stay on the ground to decompose. They will return nutrients to the soil and help retain moisture, as they do in nature. Use a lawn mower to finely chop the leaves, which will help them to decompose, dry and avoid matting or restriction of air or water. Here are some tips regarding excess dead leaves:

- Place chopped, dry leaves in the vegetable garden in the fall. The following spring, mix the decomposing leaves with the soil.
- Make compost with the dead leaves.
- Many municipalities collect dead leaves to make compost.

LAWNS

If you have decided to have a lawn, consider a low-maintenance lawn as described in Chapter 6 and use the following tips for installing and maintaining a more time-, cost- and resource-saving lawn.

Planting Methods: Seed or Sod

Most conventional lawn species, like Kentucky bluegrass, prefer sunlight and a rich, moist loam soil. If using them, it is preferable to look for the areas of your site that already possess those conditions. If you don't have them, select lawn species that are suited to the conditions you have. Table 7-3 comparing turfgrass species lists the requirements and characteristics of several species, including some that tolerate dry, infertile soils and shade, and can handle a range of soil types. Refer to Chapters 2 and 3 for tips on analyzing your site and soil.

You can amend the soil to a depth of 10-15 cm (4-6 in.) by following the tips in Chapter 3. Spread amendments, like compost, uniformly and work them into the soil without compacting the surface. Lawn species installed on the wrong kind of soil will need more maintenance later on. In shady areas, either use a shade-tolerant grass mix or try planting groundcover species.

Grass can be laid as sod or seeded. Sod is grass grown in a sod farm and then cut into strips and rolled with its soil and roots. Sodding is a bit more expensive than seeding, but it provides instant results. If the site was recently graded, seeding or sodding is best undertaken as soon as possible after grading. A delay will allow unwanted weeds to migrate to your site and

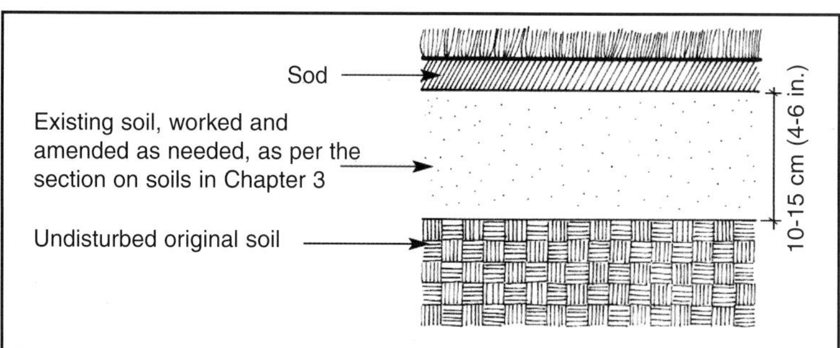

Figure 7-9: Sod planting (for slopes, refer to Figure 3-10) Installing your lawn by seed can produce a diverse lawn with a mix of species more suited to your specific property conditions. Seeded lawns are normally cheaper to install than sod, but establishment takes longer.

leave your site at risk of erosion. Installation in summer will require considerable extra watering.

Work the soil to a depth of 10-15 cm (4-6 in.). Once amendments are added and worked into the soil, smooth the surface with a rake. Create a uniform surface. Lay the sod, paying particular attention to making the sod joints tight. Alternate the joints and lay the sod perpendicular to slopes for optimal results. As you lay the sod, cut it to fit the shape of the space. Roll the lawn in one direction and then in the other. Water generously so that soil is damp to a depth of 5 cm (2 in.). Avoid walking on the sod for the first few weeks, especially when it is wet. It takes approximately 3 weeks for the sod to establish itself. Unless it rains, water daily during this time.

Installing your lawn by seed can allow more flexibility in the species mix, because you can customize a mix of species more suited to your yard's soil, moisture and sunlight conditions. Seed mixes for shady sites or that require less mowing (slow or low growing) and less watering are commercially available and may also be available in sod form. A lawn mix that is diverse is also more resistant to pests, as are ones that include fescues. Some mixes combine grass and wear-tolerant broadleaf species, like clover, that fix nitrogen, tolerate drought and poor soils and are low-growing. These are characteristics typical of low-maintenance lawns. Refer to Table 7-3 comparing turfgrass species for suitable low-maintenance lawn species. When creating your mix, there is no optimal formula, but one guideline is to include 40 per cent grasses, 40 per cent fescues and 20 per cent broadleaf species. Adjust your mix according to your site and intended use, for example, in a lawn area where you anticipate more intensive use, reduce the proportion of less durable species, like fescues.

Before seeding, follow the same soil preparation steps as described above for sod. Seed in the spring or fall but the best time is early fall because the warm soil speeds germination, roots have time to set before winter freezing, and fall rainfall can reduce the need for extra watering. If seeding by hand, scatter half the seeds while walking in one direction, then the rest in the other. You can mix the seeds with sand to help broadcast evenly. Spread them uniformly at a rate as per label instructions. Incorporate the seeds with a rake and gently roll the area to work the seeds in no deeper than 1 cm (3/8 in.). There are also mechanical seeders that bury and cover the seeds in a single operation. A light mulch, like straw, will help to keep moisture in and the seeds from scattering. Loose straw can contain weed seeds and is subject to blowing, so if possible use a straw blanket. Unless it rains, water daily for 3 to 6 weeks, but do not apply water with high force. Avoid walking in the area during this period. Don't mow the lawn until it's more than 6 cm (2½ in.) tall.

To convert an existing lawn to a low-maintenance lawn, you can stop your regular maintenance activities, then overseed with a mix of low-maintenance species, following the overseeding tips described below.

Other Lawn Alternatives

How to grow moss

If you have a moist, shady site, collect a small amount of moss from a similar location elsewhere, finely chop it and mix it thoroughly with plain yoghurt. The yoghurt acts as a growing medium and is an acidic environment that promotes moss growth. Spread this mixture on the ground or on rocks using a spatula or a rag. To help the moss

Table 7-3: Comparison of Turfgrass Species

GRASS TYPE	ZONE	DROUGHT TOLERANCE	DURA-BILITY	pH	SOIL TYPE*	WORTH NOTING
Blue Grama(W)/ (*Bouteloua gracilis*)	4-8	excellent	above average	slightly alkaline	sandy loam to clay loam	- excellent for full-sun exposed lawns and to control erosion on slopes - does not tolerate shade or heavy winter foot traffic - does not need frequent mowing, watering or fertilizing
Buffalo Grass(W)/ (*Buchloe dactyloides*)	4-8	excellent	average	slightly alkaline	sandy loam to clay loam	- drought tolerant - full sun - does not tolerate high humidity—avoid frequent irrigation - a short grass—does not need frequent mowing, watering or fertilizing
Canada Blue Grass(C)/(*Poa compressa*)	2-7	good	good	range	sandy loam to clay	- adapts to moist through dry soil conditions - tolerates poor soils
Chewings Fescue(C)/ (*Festuca rubra var. cummutata*)	2-7	very good	average	slightly acidic	sandy loam to clay loam	- full sun to partial shade - performs well in poor soil conditions
Creeping Red Fescue(C)/ (*Festuca rubra*)	2-7	good	average	slightly acidic to neutral	sandy loam to loam	- full sun to partial shade - performs well in poor soil conditions, does not tolerate wet, poorly-drained soils
Crested Wheat (C)/(*Agropyron cristatum*)	2-7	very good	average	slightly alkaline	sandy loam to clay loam	- for more informal applications where appearance is not a major factor - full sun to partial shade
Hard Fescue (C)/(*Festuca longifolia*)	3-8	very good	average	slightly acidic to neutral	sandy loam to clay loam	- good shade tolerance - slow growing, therefore reduced mowing requirements - do not over-fertilize - performs well in poor soil conditions
Kentucky Bluegrass(C)/ (*Poa pratensis*)	2-7	average	good	slightly acidic to neutral	silty loam to clay loam	- full sun to partial shade (less than 30 per cent shade) - some newer varieties have improved drought and shade tolerance
Sheep's Fescue/ (*Festuca ovina*)	2-7	excellent	average	neutral to slightly acidic	sandy, gravelly or loam	- shade tolerant - suited to low-fertility, poor soil conditions
Perennial Rye Grass (C)/ (*Lolium perenne*)	3-7	average	very good	neutral	sandy loam to clay loam soils of medium to high fertility	- full sun to partial shade - mix with Kentucky bluegrass and fine fescue for best appearance

(C) - Cool-season grasses: grow best when daytime temperatures are cool (spring and fall) and have reduced growth in the heat of mid-summer, possibly going dormant.
(W) - Warm-season grasses: do best in the heat of mid-summer and do not green-up until late spring and become dormant, turning brown, in fall.
* except heavy clay soils.
Adapted from *Household Guide to Water Efficiency*, CMHC, 2000

Low-maintenance lawn: Diversify and include broadleaf species

When lawns include a diverse mix of species, they have greater resistance to pests. You can mix the above-listed grass species with short, wear-tolerant broadleaf species, like birdsfoot trefoil and white clover, at about 20 per cent of the total mix. They have lower mowing, watering and fertilizing needs and are short and slow growing. They are drought tolerant (stay green in some of summers' hottest, driest weather), and will tolerate poor, infertile soils and clay. They capture free nitrogen in the soil and convert it to a usable form, so you don't need to fertilize as much. White clover and some of the above turfgrass species can be invasive. White clover can out-compete birdsfoot trefoil. It is less wear-tolerant than birdsfoot trefoil.

grow and attach itself, water it regularly for the first 3 weeks and during drought for the first year.

Perennial groundcovers

For shady areas where you don't anticipate foot traffic, like a side yard, perennial groundcovers can be a low-maintenance alternative to lawn. For foot traffic, you can put in a narrow path using materials like stone, crushed stone or wood chips among the groundcovers. Refer to the plant list at the end of this guide. For planting and maintenance information refer to the section on perennials below in Chapter 7.

Lawn Maintenance

Overseeding

The spring or fall following the lawn installation, overseed areas where the grass did not take hold properly. You can also improve an old lawn that is thin or has bare patches by overseeding it. This will help control weeds that germinate in bare soil patches. Before proceeding, be certain that weeds are removed. Loosen the soil surface with a rake, topdress with organic material, like compost, if needed, spread the seeds evenly and work them in no deeper than 1 cm (3/8 in.). Try overseeding with clover, or other low-growing, drought tolerant species appropriate for a lawn to save on maintenance. Keep the overseeded area moist during establishment.

Mowing:

- Set your mower blade so that you are mowing your grass to a height of no lower than 6-8 cm (2.5-3 in.). This helps shade the roots and makes the grass better able to hold water and nutrients and less susceptible to pests and disease. Cutting it too short opens it up to weed invasions and weakens the root system, making it more susceptible to heat and drought stress and disease injury. Avoiding cutting it too short will save you time, since you may be mowing less frequently.
- No more than 1/3 of the grass blade should be removed in a single mowing.
- It's best to mow when the lawn is dry but when it's not too hot and sunny.
- Each time you mow, alternate the direction to avoid grass clipping build up.
- Leave grass clippings on the lawn to add nutrients and restore moisture. Using a mulching mower will help chop the clippings finely to avoid matting and improve movement of air and water. If you do collect grass clippings, compost them rather than throwing them out with the garbage.
- Use a push mower to reduce gasoline or electricity consumption. It's also less noisy.
- Keep the blades of your mower sharp; dull blades tear the grass, leaving it open to disease and heat stress.
- Remember that a lawn composed of slow- and low-growing grasses and wear-tolerant broadleaf species requires less frequent mowing.

Watering

Established lawns generally don't need more than about 2.5 cm (1 in.) of water per week to thrive, although actual water requirements depend on factors such as soil type and species. Newly seeded or sodded lawns have greater water demands. (Refer to the section on "Planting Methods" above.) If rainfall is providing this amount of water, there is usually no need for supplemental watering. When rainfall does not provide adequate moisture, your grass may start to turn brown. This does not mean it is dead—it's simply dormant. An established lawn will recover and resume its green appearance shortly after sufficient rainfall returns.

If you want a green lawn and rainfall is insufficient to achieve that look, apply these tips to save water and money without compromising the health of your lawn:

- At the first sign of drying or discoloration, apply about 2.5 cm (1 in.) of water. If you are not comfortable with the "lookout" method, apply about 2.5 cm not more than once per week and skip a week after a good rain.The correct amount can be estimated by placing empty containers, like tuna cans, on your lawn as you apply water evenly across the surface with a sprinkler. When the water level in the containers reaches 2.5 cm (1 in.), you will know how long it takes to reach this level. Then you can set the timer on your irrigation system accordingly. If you are leaving the containers out to measure rainfall over the season, empty them once per week.

- Don't water your lawn excessively. When it's waterlogged, it may turn yellow and develop fungus and disease. Oxygen and mineral uptake may be restricted on heavy clay soils. Too much watering can also lead to thatch and fertilizer leaching.
- Avoid unnecessary traffic on your lawn and don't mow when it's dry or dormant.
- Aerate your lawn to improve water penetration. Afterwards, topdress by applying a thin layer—max. 1.5 cm (5/8 in.)—of organic material, like compost, and rake to distribute it evenly. You can overseed after this to help thicken the lawn.
- Remember that a lawn composed of species that are suited to the climate conditions of your region and to your site conditions is a water saver.
- Follow the general watering tips provided earlier in this chapter and the section comparing irrigation systems in Chapter 4.

Newly installed or seeded plants require more water than established plants. Ensuring they receive enough moisture during the establishment period is critical to their long-term health. While establishment times can vary depending on the species, climate and other conditions, follow the post-planting watering tips in the "Planting Methods" sections in this chapter.

Fertilizer

Application rates, sources and timing of fertilizer will depend on many factors including soil type and species. Knowing your site and grass will help you determine what fertilizers to use and when. Nitrogen helps your grass grow and makes it dark green, phosphorus helps the roots grow, and potassium is good for stress resistance. The numbers on fertilizer bags represent the ratio of these nutrients by weight. For example, for 12-9-9 fertilizer, 12 per cent of the total weight is nitrogen, 9 per cent is phosphorus and 9 per cent is potassium.

One rule of thumb is to fertilize once a year in the spring with a nitrogen-rich fertilizer at a rate of 1/2 kg (1 lb.) of nitrogen per 100 sq. m (1,000 sq. ft.). However, a lawn composed of species that require less fertile conditions can reduce or eliminate your need to fertilize, particularly if it includes nitrogen-fixing legumes, like white clover. Leaving grass clippings on the lawn will reduce or eliminate the need to add more nitrogen.

Natural fertilizers include manure, fish emulsion, blood meal and compost. Compost is a great source of nutrients that you can rake over the lawn. Most synthetic lawn fertilizers contain water soluble nitrogen, which is readily leachable and can be carried away, resulting in the need for regular reapplication. Most natural fertilizers contain nitrogen that does not readily dissolve in water, reducing the leaching of fertilizers into the soil and water table and releasing nitrogen at a slow rate that the plants can better utilize. They do not need to be applied as often as water soluble sources and also have a lower leaf burn potential. Refer to Chapter 3 for more details on soil analysis and amendments, including nitrogen, potassium and phosphorus.

The best conditions for fertilizing are when rain is forecast, soon after aeration and when there is low wind. Be sure to avoid fertilizing during times of drought. It is important not to over-fertilize your lawn because too much fertilizer causes your grass to grow beyond the limits of the soil and watering regime. The more fertilizer, the more water and mowing required. Excess fertilizer can also burn plants and leach into groundwater or travel via surface runoff, leading to over-nutrification of local water bodies.

Just because a weed or other pest appears, doesn't mean it's a problem. Some colonizers can become a desirable component of a low-maintenance lawn. Undesirable colonizers might include thistle, ragweed, nettle, plants that grow tall too quickly to allow for reduced mowing, or others that you personally find unsuitable. Establish your own level of tolerance for the ones you are willing to accept and those you prefer to deal with. Try to live with some dandelions or other colonizers or remove them by hand. Pull weeds when they are immature, before they go to seed. They are easier to pull when they are wet, after a rain, or a watering. Try to get it out with the whole root intact. Noxious weeds should be removed but some weeds once considered noxious are integral to native plant communities. There are some discussions underway to reconcile the growing trend towards naturalization with the need to control certain noxious weeds. If in doubt about the status of a particular plant, speak with your local weed inspector.

Weeds and Other Pests

- Follow the integrated pest management tips described earlier in this chapter, including prevention, monitoring, proper identification, and selecting the least harmful treatments. Also refer to Table 7-1 on treatments for lawn pests, like chinch bugs and white grubs.
- Lawns composed of a diverse mix of species are better able to cope with diseases and pests because if one species is attacked, the whole lawn is not affected.
- For common lawn weeds, such as crabgrass and dandelions, hand pick them with the whole root before going to seed. They are easier to pull when wet.
- Mow grass to a height of no lower than 6-8 cm (2-3 in.) so that soil is less exposed and attractive to weeds.This will also help make the grass less susceptible to disease, heat and drought stress. A thick, healthy lawn is the best prevention against weeds, insects and diseases. Bare patches or areas of heavy foot traffic are the most susceptible to weed invasions. Look for bare patches of lawn and work in some compost then overseed them to fill them in. Topdressing with compost can also help reduce susceptibility to disease.

Dethatching and Aeration

Thatch is a layer of partially decomposed grass and leaves between the soil and live grass. Some thatch is good (insulates soil, cushions the grass from foot traffic) but too much thatch can reduce water penetration and block the movement of oxygen and nutrients to the soil. If it is over 1.5 cm (5/8 in.) thick, you may need to dethatch. You can rake it in the fall or rent a thatch-remover from a garden centre. Some systems can be installed on mowers to do this.

Figure 7-10: Aerating your lawn once a year or as needed reduces soil compaction and helps organic matter, air and water reach the roots. Photo: JMK Imag-ination

Aeration helps to reduce soil compaction and provide more oxygen, water and nutrients to the roots by removing small plugs of soil and turf. Earthworms are nature's aerator—they constantly stir the soil. Aeration can be done on an as needed or annual basis in the early spring or fall. You can rent a gas-powered aerating machine and arrange with your neighbours to share the cost of renting it, or use a hand aerator.

SHRUBS AND HERBACEOUS FLOWER BEDS

Planting Methods

Select plants that are suited to your site conditions. Consult plant catalogues, ask at the garden centre or follow the directions on the tags for soil preference, sunlight requirements, height, spread and bloom time. With careful selection, your garden can bloom from early spring to late fall. Also look for hardiness zone, leaf colour and interesting fruit or bark.

Bed preparation

First, define the edge of your planting bed. Use a rope or garden hose to temporarily define the edge until you have the desired shape. (Figure 7-11) For straight lines use stakes and string to lay out the edge.

Figure 7-11: Using a garden hose to lay out the configuration of planting beds is a good way to see the bed's size and shape in real life and adjust it until you have a configuration that works. Use pegs and string to help lay out a more rectilinear design.

If you have a lawn in the area you just defined, you'll need to completely remove the grass and weeds, including the roots, which will help cut back on weeding needed in later years. Using a sod cutter, garden spade or edging tool cut out the lawn, roots and all. Shake soil off the roots. Alternatively, you can also spread about 12 layers of newspaper or a plastic sheet on the surface, after mowing the lawn close to the ground, and anchor it with rocks or soil. You'll suffocate the grass and other plants, so they won't come up next year. This can take from several weeks to an entire season, so it is best to start the year before. Dig in the dead grass after removing the plastic or paper. Then work or turn the surface.

Testing your existing soil will familiarize you with what kind of soil you have and what amendments may be needed. To the extent possible, use the soil already in place. Refer to the analysis and amendment tips in Chapter 3. Check the plant's tag or catalogues for the soil needs of your plants. Selecting plants that are best suited to your soil will save you time and money and give the best results.

Work the soil in the planting bed to a depth of about 30 cm (1 ft.), although the depth of the bed should vary according to the height of the soil in the pot or root ball of the shrubs and perennials you are planting. One common soil amendment is to work the existing soil, using a shovel (or rototiller for a large area), and add a 5-7.5 cm (2-3 in.) layer of organic material, like well-rotted compost. Work it into the soil. But the extent of the amendments needed will depend on the quality of your existing soil.

Edging plant beds

To cut down on the chore of hand edging the bed, reduce the spread of the grass and maintain the shape of the plant bed, many people will opt for hard edging. Materials can include plastic, aluminium and concrete. Refer to Chapter 5 for more information.

Some edging materials use anchoring rods to ensure they remain secure. In regions where

Figure 7-12: In your planting beds, combine species to take advantage of each other's benefits. For example, plant spring-flowering bulbs among perennials, like hosta, whose leaves emerge late, after the bulbs bloom. Once the aging leaves and stems of the bulbs fade and get pulled, the hostas emerge, filling the gap left by the bulbs. Plant species that attract beneficial birds and insects to reduce pest problems. To get continuous bloom throughout the season, plant your beds with a minimum of 12 different species with various bloom periods.

ground freezes, some edging may be subject to lifting by the frost. If the edging is not properly secured in the ground, it may move during the winter and will not fall back into place in the spring, so repair or reinstallation will be needed.

Planting shrubs and perennials

In general, the best time to plant is fall or spring. Just before planting, temporarily place potted plants directly on the bed you have prepared to give yourself a better idea of the best arrangement for the plants. You can adjust the locations according to the planting distances recommended on the tag or in the plant catalogue.

Dig holes at least twice the width of the root ball and as deep as the soil in the pot. Water the rootball before planting. When the rootball is placed on the bottom of the hole, the top of the soil around the rootball should be level with the existing soil.

After removing the plastic or peat pot by lightly tapping, cutting, pressing or turning them upside down until the undamaged root ball is free, place the plant in the planting hole. Fill the hole in layers, tamping lightly with your foot as the soil is being added. Once the existing soil level is reached, shape a water-holding saucer for the shrubs but not for the perennials. The inner edge of the saucer should correspond to the outer edge of the root ball. Slowly fill the saucer with water, allow it to penetrate into the soil and water again, allowing it to penetrate deeply into the soil.

Planting distances

When you buy plants or bulbs, they may have labels or tags indicating their characteristics including plant spacing. Sometimes it is hard to believe that a plant sold in a 10 cm (4 in.) pot or a flat will reach the dimensions indicated on the label. Yet even if this may take time, in the end it's best to follow these spacing recommendations.

Add a root stimulant like bone meal or mycorrhizae to encourage rooting and plant growth. Prune damaged or broken branches. Cover the surface with a 5-7.5 cm (2-3 in.) thick layer of mulch, being careful to pull it away from the stems of the shrubs. Refer to Table 7-2 earlier in this chapter for more details on mulch. During the first growing season, keep the soil moist until the plant is established. One rule of thumb is to water with a one-hour trickle at least once a week using a soaker hose or drip irrigation for the first three weeks after planting, barring good rainfall, and subsequently during hot, dry weather. Check soil moisture periodically, particularly during a drought and water abundantly (slowly and deeply) if it is dry.

When purchasing perennial and annual plants at the garden centre look for compact ones with healthy leaves, instead of tall lanky ones.

Examine the undersides of leaves for pests and diseases. Ask if the plants have been hardened off (gradually acclimatized from the greenhouse to the outdoors).

Annuals

Many annuals are planted at the beginning of their flowering season and removed at the end. While they have the advantage of continuous bloom throughout this period, they do not survive the winter. So, unless they are self-seeding, you will need to replant them every year, which can be costly and time consuming. The planting procedure is similar to the one described above for perennials.

To save time, money and energy purchasing, transporting and installing them every year, use annuals that will reseed themselves naturally every year. These annual flowers will reappear every year where the seeds have fallen. For details on seeding annuals or perennials, refer to the section on installing wildflower meadows later in this chapter.

Bulbs

There are non-hardy bulbs and spring-flowering bulbs. Non-hardy bulbs, like dahlias and gladiolas, do not survive frost. They must therefore be planted in the spring, after the danger of frost is past and are removed in the fall, stored and replanted in the spring after frost.

Spring flowering or perennial bulbs, like crocuses, tulips and daffodils, are planted from the end of summer until frost. They spend each winter in

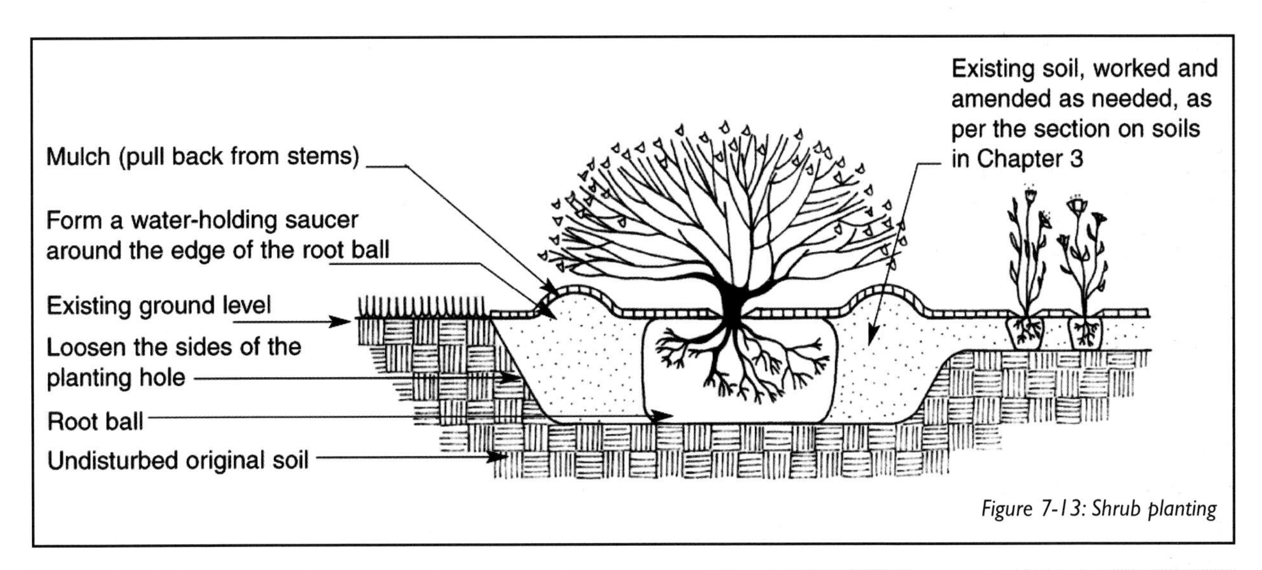

Figure 7-13: Shrub planting

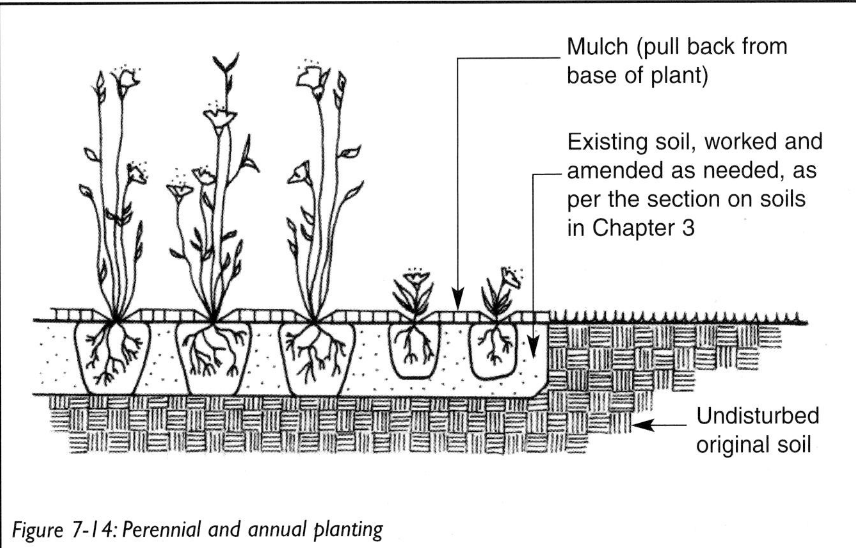

Figure 7-14: Perennial and annual planting

To save money on buying new plants, sow your own seeds or take cuttings from plants on your property that can be propagated this way. Talk to your neighbours about exchanging plants, cuttings or seeds or join a garden club. Tips on seeding are provided in the wildflower meadow section of this chapter.

the ground and will flower in the spring. (Figure 7-15) This saves you time when compared to non-hardy bulbs, that require digging up and replanting. Try naturalizing bulbs that multiply on their own in areas where you want them to spread. This will also save time and costs since they do not need to be dug up, divided or replanted.

Figure 7-15: Perennial bulbs like these daffodils, tulips and grape hyacinths, stay in ground during the winter, which will save you time digging up and storing them every fall.

A healthy bulb should be firm. Test its firmness by touch. Bulbs are planted at different depths, according to their type, but in general plant at a depth that is two or three times the bulb's diameter. Consult the packaging or ask at the garden centre for depths. Dig holes with a bulb planter or trowel and place the bulbs in the bottom of the hole with their roots pointing downwards (Figure 7-16). A root stimulant like bone meal placed in the hole encourages root growth and therefore better health. Cover them and lightly compact the soil. Then water, allowing the water to penetrate, and water again to deeply moisten the soil.

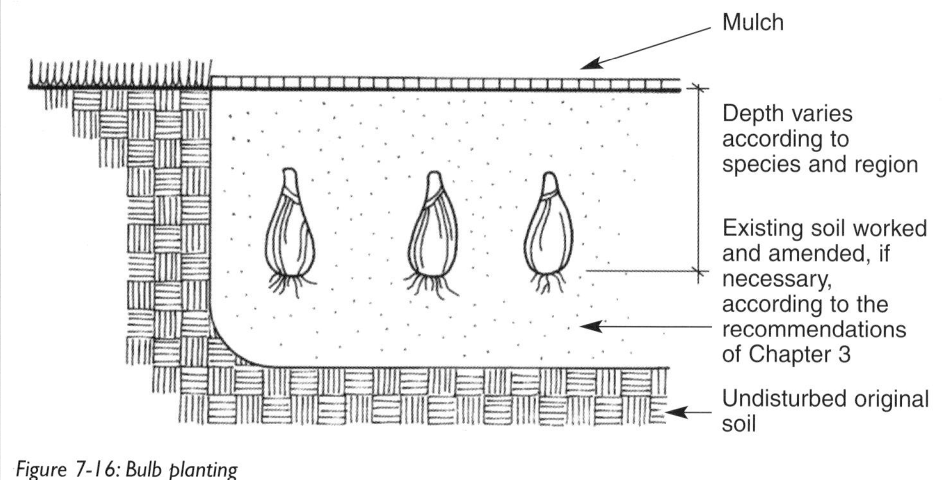

Figure 7-16: Bulb planting

Shrub/Flower-bed Maintenance

Pests and fertilization

Refer to the tips on integrated pest management earlier in this chapter, including prevention, monitoring and proper identification and using the least harmful treatment. Remember that prevention is key. For example, to control weeds, maintain a 5-7.5 cm (2-3 in.) layer of mulch.

Generally, regular fertilizer applications should not be needed if the plants you chose are well suited to your soil conditions. If not, or if you notice growth problems, you may need to fertilize. Since the fertilizer type, amount and timing depends on soil type and what plants you have, it may be best to ask your local garden centre for specific advice on specific deficiencies. Maintaining a layer of organic mulch, like compost and bark chunks, or not cutting back plants will help provide nitrogen to the plants and improve the soil.

Shrub pruning

Pruning off dead and diseased branches can be done at any time of the year. Selecting shrubs whose mature shape and size fits the location will lessen the need to prune. If you need to prune, check the type of shrub. For example, shrubs that flower in the summer set their blooms on the current year's growth and as a result, should be pruned in early spring. If they flower in the spring, they bloom on the previous year's growth, and therefore should be pruned just after they finish flowering. This will help promote the growth of young and healthy branches and abundant flowering.

You can rejuvenate old, fast-growing shrubs that are dense and have limited flowering by pruning 1/3 of the stems back to the ground. This will help let light penetrate into the centre of the plant and encourages new branches that will flower. New shoots should grow back that season.

Follow other pruning tips under the section on trees.

Weeding

A dense, healthy plant cover and good layer of mulch will go a long way to keeping down the weeds. Depending on the type, normally a layer of mulch 5-7.5 cm (2-3 in.) deep is suitable. Too much mulch can inhibit gas exchange. Since organic mulches decompose, add mulch regularly to ensure that it is an adequate depth to control the weeds. Pull weeds when they are immature, before they go to seed. They are easier to pull when they are wet, after a rain, or a watering. Try to get it out with the whole root intact.

Deadheading and cutting back

Removing dead flowers stimulates the plant to produce new flowers since it does not have to direct its energy into seed formation. This is done for aesthetic reasons. Cutting back the perennial stems is also a question of your personal taste and the time commitment. If you want a tidy look and cut back your perennials, wait until spring to cut back the ones that are attractive in the fall and winter, like ornamental grasses or perennials with interesting dried flowers. Compost the cuttings. Avoiding cutting back has the advantage of putting nitrogen back into the soil, but the dead material may become a host for insects and mold.

Dividing, separating and other propagation techniques

If they are becoming crowded or you want to propagate them, many perennials can be divided in late fall or early spring. Gently dig and lift them out, shake or brush off the soil so you can see the roots, grab a section and pull gently or cut with a sharp knife. Cut back the top and plant it at the same height as it was before. Tubers, tuberous roots and rhizomes can also be divided.

Many bulbs develop offsets that can be separated from the original. When bulbs become so crowded that flowering starts to diminish, dig and separate them. Lift a clump with a trowel or fork, pull new bulbs free and replant them.

Other propagation techniques include stem, leaf and root cuttings. Many publications are available that detail how to propagate landscape plants and list the most suitable plants.

Watering

Follow the general watering tips listed at the beginning of this chapter. For example, by applying a layer of mulch over the surface of the soil, you can reduce your water needs.

Plants should get enough water that the soil is continuously moist the first season of growth after they are planted (see above under "Planting Methods"). Afterwards, if selected to match local rainfall and site conditions, they should require little or no water other than that provided by rainfall. If rain levels are too low, the leaves of some plants may exhibit leaf burn at the edges, while others simply wilt. If you notice these symptoms you can water deeply to help restore the plant's turgidity and vigour. Moisture requirements depend on a range of factors, including soil type and species.

Direct water to the root system. For large shrubs, the roots that take up the most water are generally located within the top 30 cm (1 ft.) of the soil and are also near and beyond the drip line (the area directly below the outer tips of the branches.) Using a soaker hose or drip irrigation placed at the base of the plants on the ground (refer to Chapter 4) will help to apply water to the soil and roots—rather than the leaves or pavements—and reduce evaporation. Deep watering is better than frequent light sprinklings.

HEDGES

Choose species that are suited to your site conditions. You can refer to specialty catalogues and inquire in garden centres, and follow the instructions on the labels regarding preferred soil characteristics, height, width, hardiness zones and information on fruits, flowers and leaves.

Refer to municipal regulations regarding hedges. Hedges are often placed on the edge of a lot or right on the border of two lots. If they are to be located on or near the property line, get agreement from your neighbours.

If you are planting bare root hedges, do so when they are dormant, in early spring or, for certain species, in fall. As described earlier, in the section on trees, take precautions when transporting bare root stock. Roots should be kept constantly moist and exposed as little as possible to the sun and wind.

Before digging, lay out the edges of the future trench using string attached to stakes. Dig the full length of the planting trench. Make it sufficiently wide and deep to allow the roots to be spread out. Prepare a firm cone-shaped mound for each shrub. Place the first shrub in the trench, gently spreading the roots over the cone-shaped mound, ensuring that the trunk flare is visible at or just above the finished level of the soil outside the hole. Apply a root stimulant. Fill the hole with soil, ensuring that the trunk flare is at or just above the soil level (for soil analysis and amendments, refer to Chapter 3)

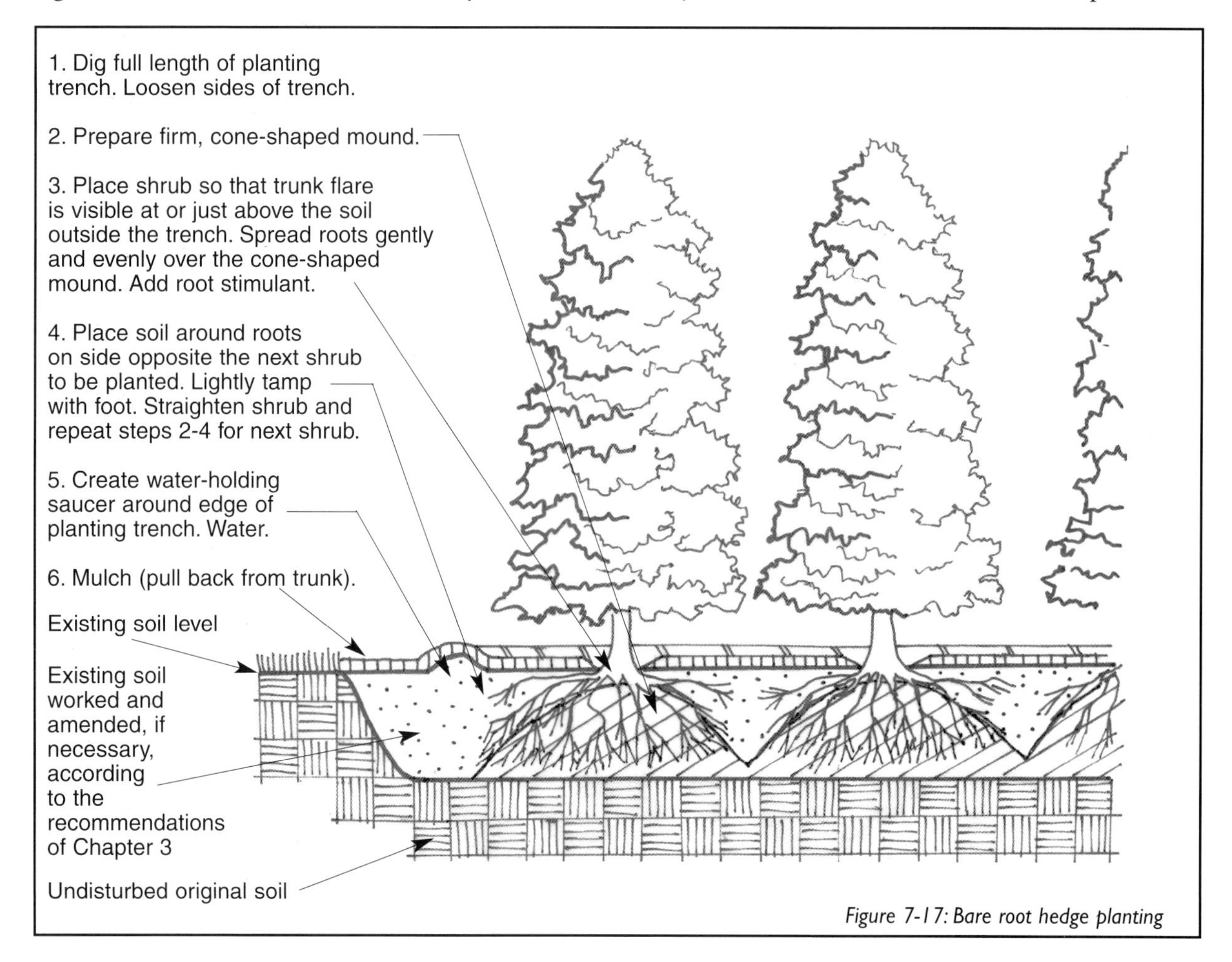

Figure 7-17: Bare root hedge planting

placing soil around the roots on the side opposite to the next shrub to be planted. Lightly tamp with your foot to avoid leaving air pockets. Straighten the plant and proceed the same way for the next shrub and so forth.

Next, form a water-holding saucer 10 cm (4 in.) high on the edges of the trench that directs the water onto the roots. Fill the saucer with water several times to water deeply and keep the soil moist with supplemental watering for a growing season or until the hedge has established. Spread a 5-7.5 cm (2-3 in.) deep layer of mulch over the saucer, being careful to pull it back from the trunks.

If you are trying to create privacy in the yard and space is limited, choose a smaller hedge species, either deciduous or evergreen. In fact, there is a much larger choice of deciduous hedges. It will be easier to find a variety that is suited to your soil, moisture and sunlight conditions. By using plants whose ultimate size and shape fits your space, you can also get a much more natural appearance and you will spend less time pruning. If you are shaping your hedge, leave the bottom wider than the top so the top growth doesn't shade the lower parts.

Refer to maintenance tips earlier in this chapter.

Figure 7-18: Wildflower meadow in the Earthwise Garden, Delta, British Columbia
Photo: Delta Recycling Society

WILDFLOWER MEADOWS AND TALL-GRASS PRAIRIES

Meadows and prairies are plant communities composed primarily of native sun-loving grasses and wildflowers. Meadows are temporary communities that are usually the result of an opening in a forest, caused by both natural and human disturbances, like fire or forest clearing for cultivation. If left alone, they will eventually evolve into woodlands, as shrubs and trees colonize them and slowly mature. Their development towards a woodland community can be stopped by burning or mowing the plants. Tall, short and mixed-grass prairies are mature, long-lived plant communities, found in certain areas of Canada, such as parts of southern Manitoba, Alberta and Saskatchewan, south-western Ontario and the inter-mountain region of B.C. Although once widespread in areas like south-western Ontario, only small fragments can be found.

Wildflower meadows and prairies are not unmown lawns gone wild that may appear neglected. They are seeded and/or planted with carefully selected species. Seeding is considerably less expensive, but the results are less predictable. There are many seed mixes available. Some mixes are composed of a high percentage of exotic plants and annuals which may result in an abundant floral show within as little as six weeks of seeding, only to rapidly fade out within a year or two as the annuals are replaced by a limited selection of tough, weedy species. Favour seed mixes with a diverse mix of native species that come from local sources or create your own mix using plants suited to your site. The development of this type of garden is a dynamic and evolutionary process. The look will change as some plants gain ground at the expense of others. More details are provided in Chapter 6.

Using the site analysis steps shown in Chapters 2 and 3, select a suitable site for your wildflower garden. They are generally best suited to sunny, dry conditions and are well suited to infertile soils.

One rule of thumb when selecting species for an attractive meadow garden is to choose at least 12 wildflower species and 4 grass species. Prairies should consist of a minimum of 8 flowers and 8 grasses. Many wildflower and grass species are shown in the plant list at the end of this guide, under the section on perennials.

Planting Method

First lay down a garden hose to delineate the area you are planting. A natural looking planting will tend to have a flowing, curvilinear shape. Completely remove the grass and weeds and their roots in the designated area by following the tips under "Bed Preparation" earlier in this chapter. This will help to eliminate competition for water, space and nutrients from the grass and weeds. Most meadow and prairie species are well suited to infertile soils, and thus do not require rich garden soil or fertilizers. After preparing the bed, smooth the surface with a rake.

Many seeds must be stratified (subjected to a period of cold) before they will germinate. Purchase them from a knowledgeable source, and inquire as to the status of the seeds that you are purchasing. If seeds have not been stratified, they should be sown in the fall and will germinate the following spring. If seeds have been stratified, spring sowing is best.

If seeding by hand, scatter half the seeds while walking in one direction, then the rest in the other. You can mix seeds with sand to help broadcast evenly. Spread them uniformly at the rate recommended on the instructions. Incorporate the seeds with a rake and roll the area to work the seeds in no deeper than 1 cm (3/8 in.). There are also mechanical seeders that bury and cover the seeds in a single operation.

Water thoroughly immediately after seeding, using a fine spray so the seeds are not washed away. Avoid walking on the surface while it is wet. Protect the seeded area with a light straw mulch, preferably a straw blanket, to keep the soil moist and keep the seeds in place. Water the seeded area daily for about three to six weeks, except on days when it rains. Keep the soil moist until the plants have established.

Figure 7-19: Seeds or small plants from pots or trays can be used for wildflower meadows. Plant in a random pattern (top). Keep soil moist until the plants establish and weed out by hand undesirable plants (middle). Mow wildflowers and tall grasses in the fall or spring (bottom).

You can also plant plugs or potted stock or combine them with seeding. In a small garden, you may want to plant the whole area with plugs or potted stock. Despite the higher upfront costs of using plugs, the success rate tends to be higher than it is for seeding. Seeding that has to be repeated for several years may be more expensive and time consuming in the long term than installing seedlings or plugs.

Arrange the plugs or pots in a natural random looking fashion, not in straight rows. Follow the information on the plant label or catalogue to know how far apart to space them. Remove the plants from their pots and place them into holes in your prepared bed twice the width of the root ball and as deep as the root ball is high. When placed in the hole, the top of the soil level around the roots should be the same as the surrounding soil. Fill the hole and tamp the soil lightly with your foot. Water thoroughly immediately after planting. Apply mulch. Keep the soil moist for at least the first three to six weeks, or until the plants have established.

Meadow/prairie Maintenance

In the first two years, weeding will be necessary but will decrease as the plants grow and spaces fill in. Little weeding should be needed after the second or third year. Regular fertilization should not be needed since the plants are well suited to infertile soils. Fertilizers generally encourage weed and turfgrass growth at the expense of your wildflower and prairie plants. Regular watering should also not be needed due to the plant's tolerance of dry conditions.

You will need to arrest the wildflower meadow's growth, for example, by mowing once per year. If not, seeds of trees and shrubs may gradually take over. Mowing can be done in the fall, but waiting until spring helps provide winter habitat and the dried plants look attractive in the winter. Some people burn their meadows every second or third year, but burns may not permitted in your area, particularly in urban areas. Check with your municipality.

WOODLAND SHADE GARDENS

Woodlands or forests are the dominant plant community throughout much of Canada, as you can see from the map shown in Figure 6-1. While they are distinguished mostly by tree cover, a complex mix of plants and creatures exist under the trees that offer ever-changing delight to any observer. More information is provided in Chapter 6.

Although woodlands take many years to establish on their own, you can speed up the process by planting young trees and nurturing them along for the first few years. The time needed to establish a woodland garden can vary widely depending on factors like the type, quantity, and size of plant material, and the presence or absence of existing trees. Sometimes, when restoring an area on the edge of an existing wood lot, you can count on existing plants to re-colonize the site. Where trees already exist, you can plant understorey shrubs and woodland wildflowers. Your intervention may be minimal, except for prohibiting access to this area to allow the plants to grow. Where no trees exist, it will take about four or more years to create a reasonably shaded woodland garden from scratch. Refer to the plant list at the end of this guide.

When selecting species, you can select a range of ages from young seedlings to two or three year old trees to larger stock. Select species that

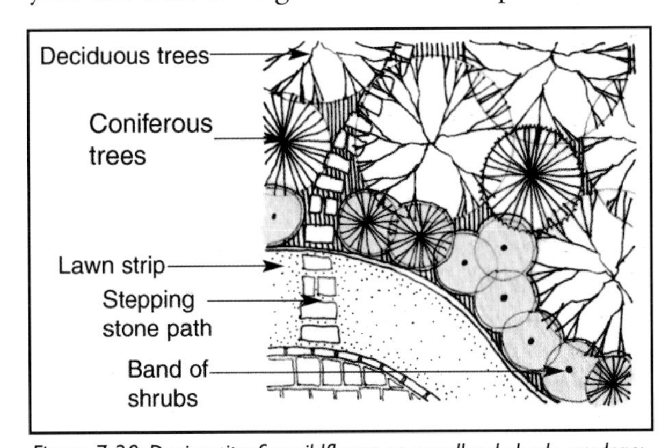

Figure 7-20: Design tips for wildflower or woodland shade gardens: Define the most visible edges with a neat band of small shrubs or perennials or leave a small area of mown grass at these edges. If neighbours have concerns that your garden will be wild or neglected, a clearly defined and neat-looking edge may help to alleviate their concerns.

are suited to the moisture conditions, soil texture, pH and fertility of your soil and favour native species. When selecting species, take clues from the native species that grow in natural wood lots in your region and emulate their arrangement. Distribute your plants irregularly, rather than in straight rows, mixing conifers with deciduous species, trees with shrubs, fast-growing with slow-growing ones and young with more mature plants.

Planting Method

First, mark out the area with a garden hose or rope on the ground. A natural looking planting will tend to have a flowing, curvilinear shape. If the area is a lawn, remove the grass and weeds and their roots by following the tips under "Bed Preparation" earlier in this chapter. This will help to eliminate competition. For larger sites where this may be too labour intensive or if you are planting larger tree and shrub stock over 1.5 m (5 ft.) in height, you may choose to not remove the lawn from the area you are planting. You can reduce competition from weeds and grass by placing a small sheet of perforated black plastic, newspaper or natural fibres around the root area of individual trees, held down by metal anchors or rocks. They will either break down or can be removed once the trees are established. Pull weeds by hand during establishment. Refer to Chapter 3 for tips on soil analysis and amendments. A common amendment is to add 5-7.5 cm (2-3 in.) of well-rotted compost.

Pioneer species, like poplar, are the fast-growing but shorter-living species that are found in the early stages of forest development growing from open, sunny meadow conditions. As their canopy growth creates shade, their fallen leaves create richer, moister soil, their root growth loosens the soil and they create the preferred conditions

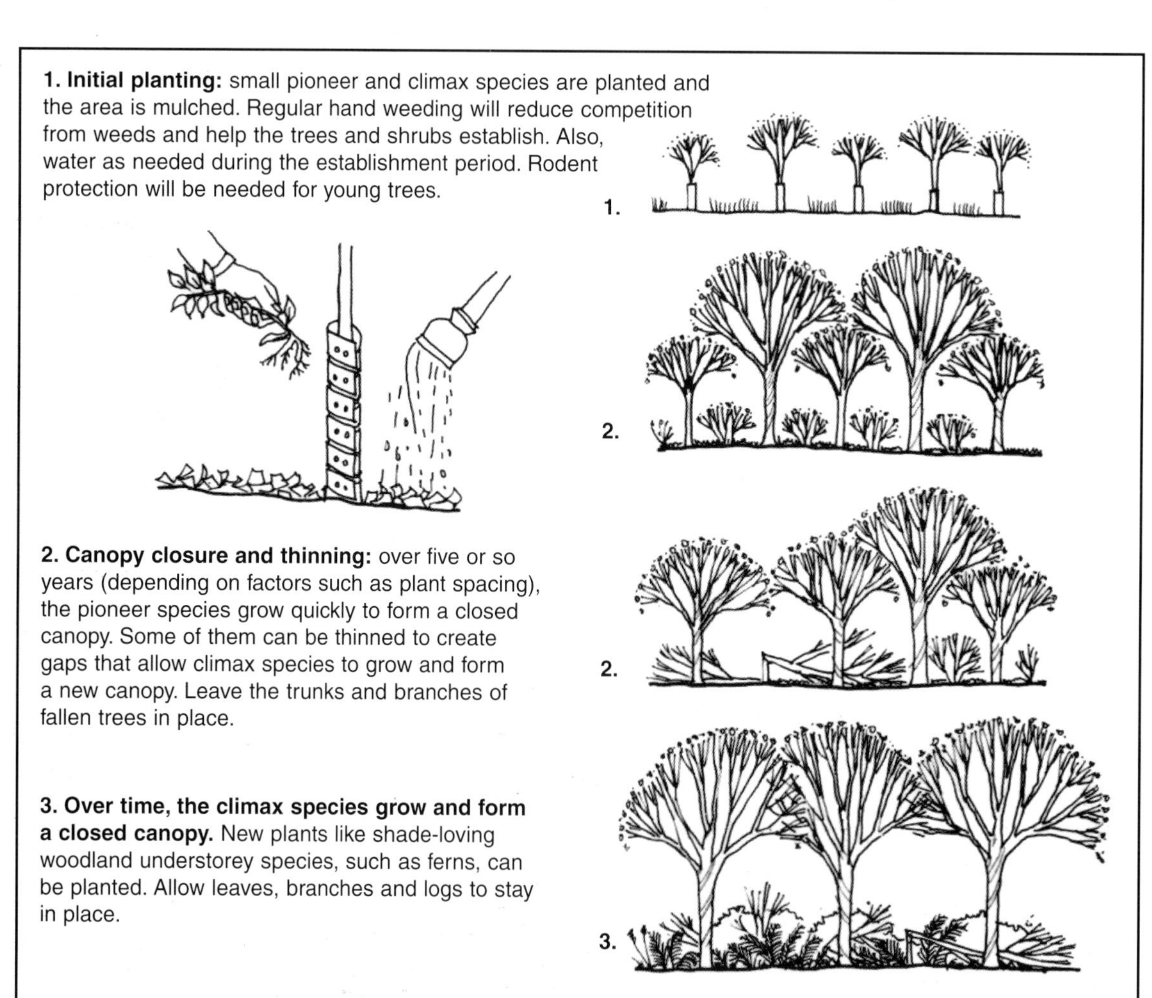

Figure 7-21: Steps in the growth of a woodland shade garden planted initially with pioneer and climax species (adapted from "Restoring Nature's Place: A Guide to Naturalizing Ontario Parks and Greenspace")

for the shade tolerant, longer-living species that are found in mature forests. Different techniques can be used to speed up the process and phase in the long-living species, depending on factors such as your budget, site conditions, proximity to a wooded area, size of the area and time-frame for achieving the desired results:

- One option is to plant only pioneer species initially, setting the stage for you to subsequently underplant with longer-living species. The pioneer species will rapidly form a closed canopy in about four to six years, depending on the spacing. At this point, you can underplant seedlings and saplings of long-lived species among the pioneers. Over the next few or more years, some pioneer species can be selectively cut (leave the trunks and branches on the ground to decompose on-site) to create gaps for the previously planted firmly rooted climax species to grow and form the new canopy. Over time additional trees, shrubs, perennials and ferns can be underplanted as needed.
- If your lot is beside an existing woodland, you could save time and costs by planting only pioneer species initially, relying on climax species to naturally colonize the site, rather than underplanting with climax species yourself some years later.
- A common approach is to combine pioneer and climax species in your initial planting (Figure 7-21). This approach helps to reduce the time and costs of subsequent thinning and replacement of pioneer species with long-lived species.
- You can bypass the pioneer stage by planting only long-lived species during the initial planting. This approach helps to reduce the time and costs of subsequent thinning and replacement of pioneer species with long-lived species. However, as these are slower-growing species, this type of planting requires more time to fill in. Also, seedlings of some species like maples, will not tolerate sun, so larger more mature stock should be used, which increases costs, and depending on the size of the area you want to plant, may be prohibitively expensive.

The spacing of the plants will depend on many factors including whether you want to have a more immediate visual impact and on your budget. Trees and shrubs may be spaced as little as 1-2 m (3-6 ft.) apart. This close spacing speeds up the formation of a canopy and helps the woodland fill in faster, but costs more initially than a wider spacing due to the higher number of plants required. To reduce costs when using this close spacing, use smaller stock, like seedlings and saplings. With this dense planting, pioneer species should be thinned later to open up the canopy to enable the other species to grow. Wider spacing about 4 m (12 ft.) apart can be used, but using larger saplings or calliper-sized stock will help the woodland establish and fill in faster. Larger stock is more tolerant of exposure and competition from other weeds, but costs more to purchase and is harder to handle during installation. The cost of using larger stock may be hard to justify for pioneer species that will be eventually thinned out.

Figure 7-22: Control weeds with mulch; watch for weeds until the tree has established. Carefully pull by hand any weeds that you do find, to eliminate competition for water, space and nutrients for the establishing tree. A weed barrier can also be used under the mulch, around the root zone of the trees. A plastic spiral bought at the nursery or a piece of corrugated drainage pipe around the base of the trunk can help to reduce damage by rodents.

Once you have considered these factors and decided on the spacing, place a marker or stake where you want trees or large shrubs. Plant in an irregular-looking staggered arrangement, rather than in a straight line. Refer to the section on trees earlier in this chapter for tips on how to plant trees.

Cover the area with a 5-7.5 cm (2-3 in.) layer of mulch, referring to the section on mulch earlier in this chapter and being careful that the mulch does not contact the base of the trunks. Again, you can help reduce competition from weeds by placing sheets of perforated black plastic, newspaper or natural fibres, held down by anchors or rocks, under the mulch immediately around the root zone of the newly planted trees. This is particularly useful on larger sites where weeding may be time consuming. You can also seed bare patches of soil with an annual grass to reduce weed colonization. Check with your nursery for suitable products. Protect young trees against rodent damage with rodent guards that can be purchased at the garden centre.

Figure 7-23: Leave fallen foliage, trunks and branches on the ground. Plant understorey species and place logs, rocks and branches to mimic the conditions of the woodland floor. Photo: Don Scallen.

Once the trees grow to a certain canopy size and there is enough shade, you can plant the understorey species. Choose hardy species that are suited to the site conditions, preferably native. Install them along the edge of the area and between the trees, referring to the section on installing shrubs and perennials earlier in this chapter.

Woodland shade gardens capture the character of natural woodlands. Placing logs, branches or rocks on the ground will help mimic the appearance and function of natural woodland. These elements will also provide habitat for birds and microfauna.

Woodland Maintenance

The end goal of this landscape option is to give part of your property back to nature. As a result, very little upkeep and watering will be required over the long term.

During the first or second growing season, the trees need to be kept moist. One rule of thumb is to water trees at least once a week with a one-hour trickle, barring good rainfall, and more frequently during hot or dry weather. Check soil moisture periodically, particularly during a drought and water abundantly (slowly and deeply) if it is dry. Once established, the need to water should be minimized by the use of plants suited to your conditions, the use of mulch and other organic material that improves soil moisture and by the protection provided by the shaded environment.

In the first years, weeding will be necessary to reduce competition. Weeds can compete with your newly installed plants for water, nutrients and air, and cause them to decline or die. Weeding will decrease as the spaces fill with plants and the tree canopy closes. Little weeding should be needed after the third year. For the most part, the shaded conditions in woodland gardens prevent the growth of most common weeds, though you do have to be wary of shade tolerant invasive species.

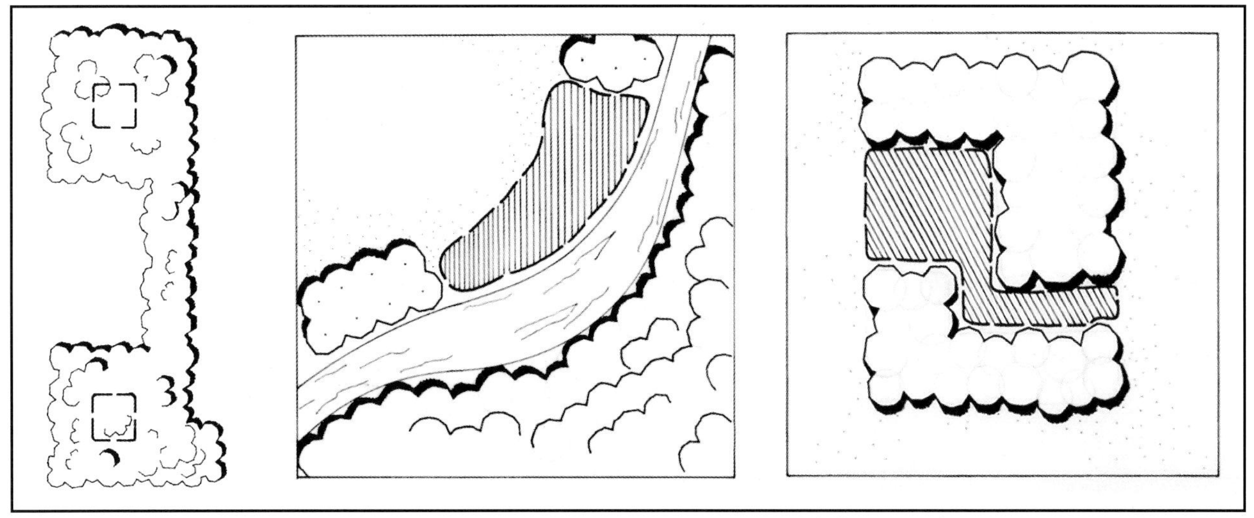

Figure 7-24: Wildlife habitat with two square forest interiors connected by a linear corridor (left, adapted from "Restoring Nature's Place: A Guide to Naturalizing Ontario's Parks and Greenspace"). To increase the size of the habitat, plant a woodland garden between separated sections of habitat. This could involve extending a linear corridor (center) or increasing the size of a woodland to enable a larger forest interior (right).

Let fallen leaves, keys, acorns, nuts etc. stay where they fall. They decompose and improve the soil. There should be no need to prune, except for dead or broken branches. If you have planted fast-growing pioneer species, after several years, some of them may need to be thinned out, as described above.

As with most of the other landscape types, actual time spent in the garden is largely dependent on each individual gardener's preferences and values. Some people may opt for a woodland garden primarily to minimize their yard work while others may enjoy the opportunity to periodically modify the garden.

Regular fertilization should not be needed if you choose plants that are suited to the soil. Also, nitrogen and organic matter are added to the soil from decaying plant material, like leaves. The broad species diversity allows the planting to better withstand pest and disease problems.

Habitat Restoration

Landowners with large properties containing wildlife habitat can preserve it and create linkages with other habitat areas. If the habitat is fragmented or separated by developed areas, the species that rely on that habitat are at higher risk. Certain species will disappear as they need a larger territory to survive and a range of conditions for food, shelter, wind and sun. You can do something to help by creating even a small habitat garden or if you have a large lot, by preserving and restoring key habitat areas and corridors, which serve as links to these areas. For example, if you live adjacent to a woodland habitat, you can increase its size by planting a woodland garden between separated sections of habitat. (Figure 7-24)

Degraded habitats can be restored. Fill in gaps in the habitat by planting species for woodland shade gardens or tall-grass prairies in the gaps, depending on the type of habitat you have. Leave native plant materials untouched, including the forest understorey, and fallen or dead branches and leaves. Remove invasive species to stop them from spreading. In these areas avoid dumping and piling materials like yard waste that can contain invasive species. Use non-invasive plants elsewhere on your property. If you live beside a body of water, refer to the section on Living by Water in Chapter 4.

Many landowners preserve and restore natural areas on their properties through conservation easements or donations to land trusts. A conservation easement is a voluntary legal agreement between the land owner and a conservation organization. It enables the land owner to preserve the conservation value on parts or all of the property, retain ownership and use of the land, establish limits on land use and receive tax benefits. Donating land to a conservation land trust also helps preserve habitat and offers considerable income tax savings.

VEGETABLE GARDENS

Vegetable gardens give back a multitude of rewards, including the satisfaction of eating food that you grew yourself. Children in particular can learn how the vegetables that they eat grow.

Pick a sunny spot. Plant rows so that tall crops are to the north and short ones are to the south with the exception of shade-loving species. These can be planted directly in the ground or in planters on your deck or balcony. Raised beds offer better drainage and make maintenance easier if you have difficulty bending over or are in a wheelchair. The soil can simply be shaped into raised rows 10-20 cm (4-8 in.) high or can be built up using rigid edging. Avoid using treated wood to edge your raised bed since the chemical products can leach into the soil.

Dig compost or other organic materials into the soil in spring. Follow instructions on your seed packet or from the garden centre on spacing of your rows, seeds and seedlings, seed depth, sunlight, soil and water requirements and thinning directions. Use organic mulch between rows to cut down on weeds. At the end of the season, clean up garden debris, so pests have no place to stay over winter or lay their eggs. Compost the debris. Turn over the soil in the garden and add compost or other organic matter like chopped dried leaves.

Particularly in small yards, make the vegetable garden a decorative element or even a centre of interest in your landscape plan. To achieve this, combine vegetables, herbs and perennial flowers. You can also incorporate structural elements, like trellises. These will also support climbing vines.

Discourage insects and diseases by rotating crops. Don't put the same vegetable in the same row next year. Also try companion planting—where one species takes advantage of the properties of another. For example, if you plant asparagus, also plant tomatoes, parsley or basil—they help control asparagus beetle. Plant certain species together that keep bugs away because of their odour or other characteristics. Certain aromatic plants like chives, dill, thyme, basil, mint, garlic, celery and onions can repel bugs. Encourage insect-eating birds to visit your garden. Plant flowers such as marigold, cosmos, portulaca and sunflower and let them go to seed.

Figure 7-25: Raised beds with narrow paths in between help to ease maintenance of vegetable gardens.

HIRING A CONTRACTOR

Chapter 8

The success of your landscape design is not based solely on good planning but also on the quality of the workmanship. As an informed homeowner, you may be able to do your own installation or parts of it. However, you may prefer to let a qualified, professional landscape contractor do some or all of the work.

To do so, it's best to go through a bidding process to get submissions or offers from contractors. This process allows you to identify the best contractor for carrying out the type of work you need done. He or she should be capable of understanding and following the plans and specifications that indicate the layout, materials and installation techniques. They should also be able to perform the work with attention to detail, while respecting the schedule and the proposed budget. Some contractors offer both design and installation services.

Many people will prefer to leave the design and preparation of technical drawings to a professional, like a landscape architect. Talk to them about the features you want and they will be able to analyze your site, prepare a design and prepare technical documents, including plans, specifications and technical cross sections, that a landscape contractor will use to do the installation. As well, they will be able to manage or watch over the site during the installation. This professional serves as an advisor and expert, whose main concern is to protect your interests.

1. Choosing a Contractor

Landscape work should be entrusted to a competent, experienced landscape contractor. There are associations in every region that check and validate the qualifications of the landscape contractors. Ensure that your contractor belongs to a trades association. For more information on landscape associations, nurseries, retailers and others associated with the landscape industry, refer to the Canadian Nursery Landscape Association Web site www.canadanursery.com Properly structured contract forms may also be available from these associations.

It is better to hire one landscape contractor to coordinate the site and if need be, have that company hire any necessary sub-contractors (irrigation, paving, stonework etc.). This contractor will then assume responsibility for proper operation on the site. He or she will ensure that all elements are carried out correctly and within set timelines.

Refer to the CMHC *About Your House* publication entitled *Hiring a Contractor* (CE 26) available at **www.cmhc.ca**

Often the best source of information is the recommendations of family, friends and neighbours who have had similar work done. The best proof of quality is satisfied customers. Hire a contractor who has already done work similar to yours. Discuss your project with a few potential contractors to get their advice and suggestions on how they would do the work. The contractors you decide to meet with should have references from at least three people they've done similar jobs for. Don't accept the references at face value. Phone them and if they'll let you, visit them to see the finished job. For a list of important questions to ask potential contractors at your initial meeting (for example, how many years have they been in business), refer to the CMHC *About Your House* publication entitled *Hiring a contractor* (CE 26).

2. Getting Estimates

Ask a minimum of three contractors to submit an estimate so that you can compare the prices and conditions. The bids submitted by contractors can be made in three ways: a fixed price, unit prices, or a cost-plus contract. With fixed-price quotes, you can be assured that all the work (except for unpredictable surprises) will be performed for the amount agreed on at the outset. In some cases allowances are established for items that you have yet to select, like a hard surface material. The allowance, which is only an estimate, is later adjusted once you have made your final material selection.

For small landscapes in particular, you can agree with the contractor on unit prices for the various materials and tasks. Once these prices are set, the contractor can carry out the proposed work, and invoice you based on quantities actually used and based on the set unit prices. With cost-plus contracts, you pay the contractor the actual cost for labour, materials and equipment and a percentage for overhead and profit. Cost-plus contracts leave costs open-ended, so it's best to set a limit so costs don't get out of hand.

Once work has started, there can be unforeseen problems due to the nature of the soil, hidden problems or other unseen conditions. Always plan a contingency of at least 10 per cent of the cost of the job to cover these eventualities.

Do not immediately choose the contractor who offers you the lowest estimate. You may have to cover additional, unforeseen costs, or you may not be satisfied with the final result. Choose the contractor who, in your opinion, will give you the most for your money. Take into account all the differences in the estimates and the contractors' respective expertise. Assume that anything not listed is not included in the price. Make sure the contractor is committed to the start and completion dates you have specified.

3. Bid Documents

For the contractor to make an estimate, you'll need to provide a copy of your landscape plans and specifications (see Chapter 2). They should clearly identify the layout, plants, grade changes and materials you are proposing in your design. The contractor will use this information to evaluate the amount of material and plants, and the complexity of their installation.

Be as specific as you can be about the materials, for example, by specifying the style and colour of precast concrete pavers. Review manufacturers brochures and catalogues to help make your selection. You may also want to specify the materials and installation methods in writing and through the use of technical cross-sections, like the ones shown in this book. You can also hire a professional to prepare these documents for you.

Your plant list should include the species, quantity of each plant, size (diameter of the trunk, height or span), spacing, and condition (bare root, potted, balled and burlapped).

Meet with the invited contractors on your property to discuss your plans and other documents to get a sense that they clearly understand the work that you want done and to see your property.

4. The Contract

Once you have accepted the estimate, the next step is to prepare a contract. For all landscape work, it is essential that you sign a clear and precise contract that clearly describes your agreement with the contractor. A contract protects both parties. The contract should specify the materials and services to be provided and installed, as well as the prices associated with them. After the contract is finalized, any changes to the work stipulated in the first contract should be made in the form of an official clause, or change order, that becomes an integral part of the contract.

The contract is an agreement that is to be signed by both parties, indicating overall price and/or unit prices as the case may be, method of payment and, sometimes, delivery deadlines. You and the contractor both sign two copies, one for each of you. A sum of money equivalent to the price of the materials may be required by the contractor on signing the contract. This common practice allows the contractor to purchase the materials in advance.

Most landscape contractors offer a one-year warrantee on the overall landscape and the plants.

The contract documents should include:

- names, addresses and phone numbers of the parties involved
- business numbers (e.g., GST, Registration number), as issued by Canada Customs and Revenue Agency
- copies of the contractor's business license (where required by municipal or provincial government) and membership in an association
- a form for change orders authorizing changes to the original plan
- a detailed description of the work, with drawings and specifications
- name of the person responsible for obtaining permits, if required, and arranging for required inspections
- total price for the job, including allowances if applicable and taxes
- payment and holdback (seasonal and lien) schedules
- hourly labour rate, for extras
- start and completion dates, with provisions for reasonable delays, and penalties
- workers Compensation number and proof of workplace and business liability insurance coverage
- list of subcontractors to be used on the job
- details of contractor's warranty and maintenance agreement
- a statement of how disputes are to be settled, with name of mediator, if appropriate
- termination agreement, specifying what it will cost you to void the contract at different stages of the work
- a statement that the contractor will provide you with manufacturer's warranties for products supplied by the contractor and used in the work

Do not sign the contract until:

- You have read it carefully.
- You understand what it says.
- You are satisfied that it describes exactly what you want and that it includes everything you have been promised.

5. Get it in Writing

Do not be tempted by a contractor who doesn't have an address, doesn't want a written contract and offers a discount if you pay cash. This type of underground economy transaction involves many risks and pitfalls that offset any savings.

A cash deal may leave you with no legal recourse if something goes wrong or the work isn't satisfactory, or if the contractor walks off the job without finishing it. In fact, it makes it difficult for you to prove the contractor was ever there. And after you have paid the contractor, you may find that materials haven't been paid for or workers haven't been paid—and you are responsible for the bills. For your own protection and peace of mind, it's best to deal in a legal and responsible way—always get it in writing.

For information on paying for the work, liens and seasonal holdbacks, completion certificates, working with your contractor, consumer protection laws and insurance, refer to the CMHC *About Your House* fact sheet CE26 on Hiring a Contractor. Also refer to the *Sample Renovation Contract*. These are available on the Internet at **www.cmhc.ca**

6. Construction Method

The extent of the supervision required during the landscape work is based just as much on the complexity of the design as on the knowledge and experience of the contractor. When several contractors work together on the same site, it is essential to maintain efficient coordination between them. If there is any doubt, you can have the work checked by an independent expert, such as a landscape architect, who will see that the work is done as planned, will check the quantity and quality of the materials installed, issue periodic payment certificates and approve necessary changes on your behalf.

Here is a list of the critical phases:

Beginning Construction

At the beginning of construction, you should get together with the designer and the contractor to study the scope of the contract and plan the work commencement. This visit can also be used as a preliminary inspection for the labour and equipment that will be used on the worksite. During this meeting, the contractor may be asked to provide a sample of the materials and colours that will be used. If it is not already in the technical documents, an agreement can be made about the source of the materials.

Protection of Existing Vegetation During Construction

The existing vegetation that you want to preserve should be protected against damage caused by construction equipment. The dollar value of a tree can be calculated by a certified arborist. You can build this dollar value into your contract, to protect yourself against damage the contractor may cause to existing trees. Refer to the section on Preserving Natural Features and Existing Plants in Chapter 6. Your contract should include tree protection methods, like fencing and restrictions within protection zones.

Erosion Prevention During Construction

Erosion during construction can have long-term repercussions. Erosion should be prevented, especially during the grading work. Here are some tips to follow during construction:

- Avoid creating steep slopes.
- Expose bare soil only for the least possible time during the construction work. Stabilize slopes temporarily with coverings, like straw.
- Strip existing topsoil off areas that will not be used for lawn, planting beds, trees and any other kind of planting. Store and reuse it on your property for plantings.

Laying Out Landscape Features

The various features in your landscape plan or the layout plan can be physically located on your property by the contractor and the designer to create consensus at the outset. This will enable you to see the size, shape and arrangement of the various features of your design before the installation begins.

Installing Hard Materials

A close inspection should be performed at the time of installation of underground elements, like the foundation under pavements, as once the work is completed, it is often impossible to check the condition of the foundations.

Work Completion

The final visit to the worksite is usually the last meeting between you, the supervisor (if you have one) and the contractor, the time at which the finished work is accepted. If the work has been performed in accordance with the contractual documents and if the parties are satisfied with the outcome, the balance of the contractor's payment is then paid. It is recommended that the inspections be performed at regular intervals and at the time of the critical construction phases.

Maintenance

Many homeowners look after garden maintenance themselves. Those who don't will normally hire a contractor who specializes in maintenance. The steps for getting estimates and developing a contract are similar to those outlined in this chapter.

PLANT LIST - TREES, SHRUBS, PERENNIALS AND CLIMBING PLANTS

DECIDUOUS TREES

Hardiness Zone	Latin name / Common name	Characteristics / Special visual features*	WUseful for xeriscape (X), woodland (W) meadow/ prairie (P)	Sun exposure			Moisture			Soil	Height
				❍	◗	●	♦	♦♦	♦♦♦		
6	*Acer circinatum* / Vine Maple	- from southern B.C. - attractive fall colours	W	❍	◗			♦♦		- adaptable, but prefers moist, well-drained soil	7 m (23 ft.)
5a	*Acer glabrum douglasii* / Douglas Maple, Rocky Mountain Maple	- from B.C. - multi-stemmed, shrubby - good for small gardens - attractive fall colour	W	❍	◗			♦♦	♦♦♦	- moist, well-drained soils	10 m (33 ft.)
6	*Acer macrophyllum* / Bigleaf Maple	- from southern coastal B.C. - large leaves - hanging clusters of fragrant flowers		❍	◗			♦♦	♦♦♦	- generally grows in gravelly, moist soils but is found on other types of moist soils	30 m (100 ft.)
2b	*Acer negundo* / Manitoba Maple	- from Manitoba and Saskatchewan - colonizes disturbed sites	X, W	❍			♦	♦♦	♦♦♦	- native to stream, lake, swamp edges but does well on poor, dry or wet soils	16 m (52 ft.)
3	*Acer rubrum* / Red Maple	- from Ontario to Newfoundland - red colour in the fall	W	❍	◗			♦♦	♦♦♦	- prefers moist slightly acid soil - sand, silt, loam or clay	20 m (66 ft.)
2b	*Acer saccharinum* / Silver Maple	- from southern Quebec and Ontario - appropriate for large spaces - aggressive root system	W	❍	◗			♦♦	♦♦♦	- tolerant of wide variety of soils - sand, silt, loam or clay, - prefers moist soils	24 m (80 ft.)
4a	*Acer saccharum* / Sugar Maple	- from Ontario to Nova Scotia - yellow/orange to red colour in the fall	W	❍	◗	●		♦♦		- prefers deep, fertile, moist, well-drained soil - intolerant of compaction	20 m (66 ft.)
1a	*Betula papyrifera* / Canoe Birch, White Birch, Paper Birch	- from all forested regions of Canada - attractive in winter - interesting bark	W	❍				♦♦		- does best on well-drained sandy loam, moist soil	15-21 m (50-70 ft.)
3b	*Carpinus caroliniana* / Blue Beech	- from southern Ontario and Quebec - interesting for its globular form and twisted trunk	W		◗	●		♦♦		- deep, rich, moist, but will grow on drier sites - silt or loam to clay	8 m (26 ft.)
4a	*Carya cordiformis* / Bitternut Hickory	- from southern Quebec and Ontario	W	❍	◗			♦♦		- prefers moist, rich, well-drained loams	20 m (66 ft.)
4b	*Carya ovata* / Shagbark Hickory	- from southern Quebec and Ontario - interesting golden yellow colour of leaves in fall and shaggy bark	W	❍	◗			♦♦		- adaptable, sand, silt or loam clay - prefers moist soil	23 m (75 ft.)
2b	*Celtis occidentalis* / Hackberry	- from southern Quebec and Ontario - tolerates urban conditions	X, W	❍	◗		♦	♦♦	♦♦♦	- adaptable, heavy or sandy, wet to dry but prefers rich, moist sandy loam	18 m (59 ft.)
6b	*Cornus florida* / Eastern Flowering Dogwood	- from southern Ontario - small low-branched tree, with attractive flowers, leaves, fruit and form	W		◗	●		♦♦		- well-drained, rich, acid moist soil	7 m (23 ft.)
8	*Cornus nuttallii* / Pacific Dogwood	- from Vancouver Island and southern B.C. coast - also shrub form to 10 m	W	❍	◗			♦♦	♦♦♦	- well-drained loam soil	20 m (66 ft.)
2b	*Crataegus crus-galli* / Cockspur Hawthorn	- from southern Quebec and Ontario - attractive fruit - has sharp thorns - tolerant of urban conditions	W	❍	◗			♦♦		- well-drained	6 m (20 ft.)

Hardiness Zone	Latin name / Common name	Characteristics / Special visual features*	Useful for xeriscape (X), woodland (W) meadow/ prairie (P)	Sun exposure			Moisture			Soil	Height
				❍	◗	●	♦	♦♦	♦♦♦		
4a	*Fagus grandifolia* / American Beech	- from Lake Huron to Nova Scotia - appropriate for large spaces	W	❍	◗	●		♦♦		- prefers moist, acid, well-drained loam soil	22 m (72 ft.)
3b	*Fraxinus americana* / White Ash	- from Lake Huron to Nova Scotia - yellow or purple foliage in the fall - tolerant of urban conditions, disturbed soils	W	❍				♦♦		- adaptable, but prefers deep, moist, well-drained soil	20 m (66 ft.)
2b	*Fraxinus nigra* / Black Ash	- from Manitoba to Atlantic - nice colours in the fall, golden / yellow foliage	W	❍				♦♦	♦♦♦	- tolerates wet soils - sand, silt or loam, clay, organic soils	16 m (62 ft.)
2b	*Fraxinus pennsylvanica* / Red Ash, Green Ash	- from Saskatchewan to Nova Scotia - yellow fall colour	W	❍				♦♦	♦♦♦	- sand, silt or loam, clay, - tolerates wet soil - once established, tolerates salt, dry and infertile soil	18 m (59 ft.)
4	*Gingko biloba* / Maidenhair Tree	- tolerant of urban pollution, salt - bright yellow fall colour, fan shaped leaves		❍				♦♦		- prefers sandy, moderately moist soil	20 m (66 ft.)
7b	*Malus fusca* / Pacific Crab Apple	- from coastal B.C. - small tree, large shrub - fragrant white flowers, yellow or red fall colour and fruit	W	❍	◗			♦♦		- prefers well-drained, moist soil	9 m (30 ft.)
3a	*Ostrya virginiana* / Ironwood	- from eastern Manitoba to Nova Scotia - appropriate for small shady areas - tolerates urban pollution	W		◗	●	♦	♦♦		- sand, silt or loam, clay	12 m (40 ft.)
1a	*Populus balsamifera* / Balsam Poplar	- from across most of Canada - pioneer plant in the boreal forest - used in rural areas for windbreaks, - suckers, spreads	W	❍				♦♦	♦♦♦	- sand, silt or loam, clay	25 m (82 ft.)
2b	*Populus deltoides var. occidentalis* / Plains Cottonwood	- from parts of prairie provinces - tolerates salt and pollution - used in rural areas for windbreaks	W	❍			♦	♦♦	♦♦♦	- sand, silt or loam, clay	28 m (92 ft.)
1b	*Populus tremuloides* / Trembling Aspen	- from across most of Canada - spreads by suckers - like other poplars its a pioneer species, fast growing, short lived, branches break easily	X, W	❍				♦♦	♦♦♦	- can be found in moist, loamy sand to rocky soil and clay	12 m (40 ft.)
7b	*Prunus emarginata* / Bitter Cherry	- from parts of B.C. - pinkish white flowers, red bitter fruit		❍	◗			♦♦	♦♦♦	- moist, well-drained soil	15 m (50 ft.)
1a	*Prunus pennsylvanica* / Pin Cherry	- from central B.C. to Eastern Canada - its fruit attracts birds - pioneer species	X, W	❍			♦	♦♦		- sand, silt or loam	10 m (33 ft.)
2b	*Prunus serotina* / Black Cherry	- from southern Ontario to Nova Scotia - its fruit attracts birds - white, lightly-perfumed flowers in spring	W	❍			♦	♦♦		- sand, silt or loam	20 m (66 ft.)

2a	*Prunus virginiana* / Choke Cherry	- from central B.C. to Newfoundland - its fruit attracts birds - white perfumed flowers in spring - pioneer species	W	○	◗		💧	💧💧		- adaptable but prefers well-drained, rich, moist soil	9 m (30 ft.)
4a	*Quercus alba* / White Oak	- from southern Quebec and parts of Ontario - its fruit attracts birds and mammals - red/violet foliage in the fall	W	○	◗		💧	💧💧		- adaptable, but prefers moist, well-drained soil	25 m (82 ft.)
4b	*Quercus macrocarpa* / Bur Oak	- from south-eastern Saskatchewan to parts of New Brunswick - its fruit attracts birds and mammals - brown leaves in fall	W	○			💧	💧💧	💧💧💧	- very adaptable, sandy to clay soils, dry to wet	25 m (82 ft.)
3a	*Quercus rubra* / Red Oak	- from Lake Superior to Nova Scotia - its fruit attracts birds and mammals - red foliage in fall - withstands urban pollution	W	○			💧	💧💧		- prefers well-drained sandy loam, neutral to acid	24 m (79 ft.)
2a	*Salix spp.* / Willows	- many species, range in size from dwarf (see shrubs below) to 20+ m e.g. black willow - usually fast growing, extensive roots, branches break easily - often used for restoring wetlands, river banks and lakeshores	W	○				💧💧	💧💧💧	- range of soil types depending on species - usually moist to wet soil	6-20 m (20-66 ft.)
2a	*Sorbus decora* / Showy Mountain Ash	- from central Manitoba to Newfoundland - interesting for its orange-red fruit that attracts birds - small tree or large shrub, similar to *Sorbus Americana*/ American Mountain Ash	W	○				💧💧	💧💧💧	- sand, silt or loam, clay	7 m (23 ft.)
3a	*Tilia americana* / Basswood	- from southern Manitoba to southern New Brunswick - highly perfumed yellow flowers	W	○	◗	●		💧💧		- sand, silt, loam, clay, but prefers fertile, deep moist soil	23 m (75 ft.)
3	*Tilia cordata* / Littleleaf Linden	- tolerates urban pollution - fragrant flowers		○				💧💧		- prefers moist, well-drained, fertile soil	12 m (40 ft.)
CONIFERS											
1a	*Abies balsamea* / Balsam Fir	- from northern forests of central and eastern Canada - used for Christmas trees	W	○	◗	●		💧💧	💧💧💧	- prefers well-drained, acid, moist soil - sandy to loam	20 m (66 ft.)
2b	*Abies lasiocarpa* / Subalpine Fir	- from Yukon to western Alberta and B.C. - high altitude tree	W	○	◗	●	💧	💧💧		- found on a variety of soils, but does best on well-drained loam	30 m (100 ft.)
4	*Abies concolor* / White Fir	- tolerates drought	X	○	◗		💧	💧💧		- prefers moist, well-drained, rich soil, but tolerates dry	15 m (50 ft.)
8	*Arbutus menziessii* / Madrone	- from southern coastal B.C. - broadleaf evergreen - peeling brownish red bark shows yellow underbark	W	○			💧			- well-drained soil	10-20 m (33-66 ft.)
2b	*Juniperus communis* / Common Juniper	- from across most of Canada - cultivars with different forms and sizes are available - tolerant of windy sites	X	○			💧			- grows on dry, sandy, infertile, rocky, poor sites	3 m (10 ft.)

Hardi-ness Zone	Latin name / Common name	Characteristics / Special visual features*	Useful for xeriscape (X), woodland (W) meadow/ prairie (P)	Sun exposure			Moisture			Soil	Height
				❍	◗	●	♦	♦♦	♦♦♦		
2a	*Juniperus horizontalis* / Creeping Juniper	- from across most of Canada - tolerant of windy sites	X	❍			♦			- grows on sandy, rocky soil - tolerant of dry sites	60 cm (2 ft.)
3	Juniperus *scopulorum* / Rocky Mountain Juniper	- from parts of B.C. - drought tolerant	X	❍			♦	♦♦		- adaptable, dry, sandy soils	10 m (33 ft.)
3a	*Juniperus virginiana* / Eastern Red Cedar	- from parts of Ontario and west Quebec - tolerant of difficult sites	X	❍			♦	♦♦		- tolerates poor, dry, gravely sites, prefers moist, well-drained loam	12 m (40 ft.)
1	*Larix laricina* / Tamarack, Eastern Larch	- from eastern Yukon to Newfoundland - golden colour in fall - loses its needles in the fall	W	❍				♦♦	♦♦♦	- grows best in moist, acid, well-drained soil but also found mostly on wet, poorly drained sites	20 m (66 ft.)
1	*Picea glauca* / White Spruce	- from all forested regions of Canada except coastal B.C. - withstands drought	W	❍	◗			♦♦		- grows on a variety of soils, prefers rich, moist soil	22 m (72 ft.)
1a	*Picea mariana* / Black Spruce	- from northern forests from Yukon to Newfoundland	W	❍	◗			♦♦	♦♦♦	- grows on a variety of sites, usually wet to moist, poorly-drained organic soils	12 m (40 ft.)
2	*Picea pungens* / Colorado Spruce	- cultivars with bluish colour, some shrub forms	W	❍	◗			♦♦		- adaptable, prefers rich, moist soil	20 m (66 ft.)
1a	*Pinus banksiana* / Jack Pine	- from northern forests of Canada, from the North West Territories and Alberta to Nova Scotia - extremely hardy	X, W	❍			♦	♦♦		- poor, dry, acidic soils: course sand, shallow, rock outcroppings	16 m (52 ft.)
3a	*Pinus contorta var.contorta* / Shore Pine	- from coastal B.C. - twisted, crooked form - also *Pinus contorta latifolia*/ Lodgepole Pine (from southern Yukon to Alberta)	X	❍			♦	♦♦	♦♦♦	- grows in nutrient-poor soil - found on rocky ridges, coastal sand dunes and in bogs	to 20 m (66 ft.)
4	*Pinus nigra* / Austrian Pine	- tolerates urban conditions: air pollution, salt spray, dry soil - attractive form, dark green needles		❍			♦	♦♦	♦♦♦	- adaptable, prefers moist, well-drained soil but tolerates dry and clay soil	18 m (59 ft.)
2b	*Pinus ponderosa* / Ponderosa Pine	- from parts of southern B.C. - drought tolerant	W	❍				♦♦		- grows on a variety of soils	18-60 m (60-200 ft.)
2b	Pinus resinosa / Red Pine	- from south-east Manitoba to Newfoundland - reddish, scaly bark	X, W	❍			♦			- does well on sandy, dry, acid soils	24 m (79 ft.)
2b	*Pinus strobus* / Eastern White Pine	- from south-east Manitoba to Newfoundland - sensitive to salt spray	W	❍	◗			♦♦		- prefers fertile, moist, acidic, sandy loam soil	23 m (75 ft.)
7b	*Pseudotsuga menziesii* / Douglas Fir	- from Rocky Mountains and Pacific Coast - one of the largest of Canada's native conifers		❍				♦♦		- prefers deep, well-drained, moist loams	12-60 m (40-200 ft.)
3a	*Taxus canadensis* / Canada Yew	- from Newfoundland to Manitoba - low-spreading shrub	W		◗	●		♦♦		- moist, sandy loam	1 m (3 ft.)

6	*Thuja plicata* / Western Red Cedar	- from parts of B.C. - shaggy, reddish bark, buttressed base	W	❍	◗			♦♦		- prefers moist, well-drained fertile soil	15-60 m (50-200 ft.)
3	*Thuja occidentalis* / Eastern White Cedar	- from south-east Manitoba to Nova Scotia - common as a hedge plant	W	❍	◗			♦♦	♦♦♦	- sand, silt, loam, clay, neutral to alkaline - moist to wet	15 m (50 ft.)
4a	*Tsuga canadensis* / Eastern Hemlock	- from southern Ontario to Nova Scotia - plant in sheltered location - avoid windy, polluted sites	W		◗	●		♦♦	♦♦♦	- sandy, silt or loam, moist, acid soil	20 m (66 ft.)
6	*Tsuga heterophylla* / Western Hemlock	- from B.C., coastal and mountain	W	❍	◗	●		♦♦	♦♦♦	- variety of soils, prefers moist	50 m (165 ft.)
DECIDUOUS SHRUBS											
1a	*Alnus crispa* / Green Alder	- from Labrador to Alaska and southward - useful for stream bank rehabilitation		❍	◗			♦♦	♦♦♦	- sandy, gravelly, usually moist to wet	3 m (10 ft.)
1	*Amelanchier alnifolia* / Saskatoon-berry	- from Manitoba and Saskatchewan east to parts of Ontario and Quebec - white, fragrant flowers - purple to black, sweet, edible fruit	W	❍	◗			♦♦		- sand, loam, silt	4 m (13 ft.)
3a	*Amelanchier arborea* / Downy Serviceberry or Juneberry	- from Lake Huron to southern Quebec - nice colours in fall, purplish black fruit attracts birds, white flowers in spring - similar to *Amelanchier canadensis*/ Shadblow Serviceberry	X, W	❍	◗		♦	♦♦		- adaptable, prefers moist, well-drained soils, also grows on drier sites	7 m (23 ft.)
3b	*Amelanchier laevis* / Allegheny Serviceberry	- from Newfoundland to Lake Superior - sweet, black fruit attracts birds	W		◗	●		♦♦		- adaptable, prefers moist, well-drained loams	8 m (26 ft.)
2a	*Arctostaphylos uva-ursi* / Bearberry, Kinnikinick	- from across much of Canada - good salt tolerance - a reliable groundcover	X, W	❍	◗		♦	♦♦		- adaptable, does best on poor, gravelly or sandy, infertile, acidic soils	15-30 cm (6-12 in.)
4	*Aronia melanocarpa* / Black Chokeberry	- from southern Ontario, Quebec and Maritimes - black fruit attracts birds - attractive fall foliage	X, W	❍	◗		♦	♦♦	♦♦♦	- adaptable (except compacted clay) - wet or dry	1 m (3 ft. 3 in.)
2	*Caragana arborescens* / Siberian Pea Shrub	- lacy foliage, ideal for hedges and screens, yellow flowers - easy to grow, very hardy - tolerant of drought, poor soil, salt and wind	X	❍			♦			- tolerates poor soil but well-drained	3.5 m (11½ ft.)
3b	*Cornus alternifolia* / Alternate-leaf Dogwood	- from part of southern Manitoba and Lake Superior to western Newfoundland - white flowers in spring and red leaves in fall - blue-black fruit with pinkish red stalk	W		◗	●		♦♦		- moderate to well-drained moist loam	8 m (26 ft.)
2	*Cornus canadensis* / Bunchberry	- from forests across Canada - low groundcover with white flowers and red fruits - evergreen	W		◗	●		♦♦		- moist, humus-rich, acidic soil	20 cm (8 in.)

Hardiness Zone	Latin name / Common name	Characteristics / Special visual features*	Useful for xeriscape (X), woodland (W) meadow/ prairie (P)	Sun exposure			Moisture			Soil	Height
				❍	◗	●	♦	♦♦	♦♦♦		
2b	*Cornus racemosa* / Grey Dogwood	- from southern Ontario - white berries attract birds - red leaves in fall	W	❍	◗			♦♦	♦♦♦	- grows best in moist, cool, well-drained soil but is adaptable (sand, silt, loam, clay)	3 m (10 ft.)
2a	*Cornus stolonifera* / Red Osier Dogwood	- Alaska to Labrador and Newfoundland—spreads rapidly - good for massing in large areas - red bark—attractive in winter	W	❍	◗			♦♦	♦♦♦	- very adaptable, best in moist soils	3 m (10 ft.)
1	*Corylus cornuta* / Beaked Hazel	- from B.C. to Newfoundland - tall shrub or small tree	W	❍	◗			♦♦		- sand, loam soil	2-4 m (6½-13 ft.)
3a	*Diervilla Lonicera* / Low-bush Honeysuckle	- from Newfoundland to parts of Saskatchewan - yellow to red fall colour	X, W	❍	◗	●	♦	♦♦		- adaptable, from sand to clay - tolerates dry, infertile soils	1 m (3 ft. 3 in.)
4a	*Dirca palustris* / Leatherwood	- from Lake Huron to southern New Brunswick - rich green leaves, yellow in fall	W	❍	◗	●		♦♦		- fertile, moist sand, silt, loam soil	1.5 m (5 ft.)
2	*Elaegnus commutata* / Silverberry, Wolf Willow	- from eastern Canada to Northwest Territories south to Prairies - silvery leaves - a nitrogen fixing shrub	X	❍			♦	♦♦		- poor soils, sand, silt, loam, clay	1-4 m (3-13 ft.)
7b	*Gaultheria shallon* / Salal	- from B.C. - evergreen, purple berries, red twigs - drought tolerant once established	X, W	❍	◗	●	♦	♦♦	♦♦♦	- dry to moist, nutrient poor, well-drained acidic soil	1-2 m (3-6½ ft.)
2a	*Gaultheria procumbens* / Wintergreen	- from Newfoundland to south-east Manitoba - creeping, evergreen groundcover - bright red edible berries	X, W		◗	●	♦	♦♦	♦♦♦	- best on acidic, sand, loam soil high in organic matter	10 cm (4 in.)
4b	*Hamamelis virginiana* / Witch-hazel	- large shrub from southern Ontario and southern parts of Quebec, N.B. and N.S.	W	❍	◗	●		♦♦		- found on a variety of sites, mostly moist, sandy, silt or loam soil	5 m (16 ft.)
2b	*Hydrangea arborescens* / Smooth Hydrangea	- large, white flowers -old flowers turn brown and remain on the plant in winter		❍	◗	●		♦♦		- prefers rich, porous, moist soil	1.25 m (4 ft.)
4a	*Hypericum prolificum* / Shrubby St. John's-wort	- from Quebec and Ontario - rounded, dense shrub - dark green leaves, bright yellow flowers	W	❍	◗		♦	♦♦		- well-drained, dry to moist, does well in dry soil	1 m (3 ft. 3 in.)
3b	*Ilex verticillata* / Winterberry	- from Newfoundland to eastern Lake Superior - red or orange fruit	W	❍	◗			♦♦	♦♦♦	- prefers wet to moist, sand, silt or loam, high in organic matter	90 cm (3 ft.)
3a	*Lonicera canadensis* / American Fly Honeysuckle	- from south-east Manitoba to Newfoundland	W	❍	◗	●		♦♦		- sandy to clayey loams	2 m (6½ ft.)
5	*Mahonia aquifolium*/ Oregon Grape	- from B.C. - broadleaf evergreen - holy-like, dark green leaves that turn purplish-bronze in winter - yellow flowers, blue-black fruit	W	❍	◗	●		♦♦		- prefers moist, organic, acid, cool, course to medium soil	1-1.8 m (3-6 ft.)

1b	*Myrica gale* / Sweet Gale	- from across Canada, except most southern regions	W	❍	◗			♦♦	♦♦♦	- sandy, silty, organic soil	1- 2 m (3-6½ ft.)
2b	*Physocarpus opulifolius* / Ninebark	- from Great Lakes/St. Lawrence basin - used for massing, screens	X, W	❍	◗		♦	♦♦		- well-drained soil - dry to moist	2.5 m (8 ft.)
2a	*Potentilla fruticosa* / Shrubby Cinquefoil	- from across much of Canada - flowers profusely and for a long period - tolerant of urban conditions	X	❍	◗		♦	♦♦	♦♦♦	- adaptable, dry to wet, poor, sand to clay soils but best in fertile, moist, well-drained soil	60 cm-1.3 m (2-4 ft.)
3	*Rhus typhina* / Staghorn Sumac	- from Lake Huron to Nova Scotia - bright red leaves in fall - red fruits attract birds - produces suckers—best used in masses, e.g. along banks	X	❍	◗		♦	♦♦		- prefers well-drained, tolerates very dry, sterile soil	6 m (20 ft.)
2	*Ribes alpinum* / Alpine Current	- yellow flowers - used for hedges	X	❍			♦	♦♦		- well-drained, slightly alkaline soil	1 m (3 ft. 3 in.)
2	*Rosa blanda* / Smooth Wild Rose	- from N.B. to eastern Saskatchewan - thornless stems - tolerant of urban conditions	W	❍				♦♦		- sand, silt, loam, clay	1.5 m (5 ft.)
5	*Rosa nutkana* / Common Wild Rose	- from Alaska to California, B.C. interior - pink or white flowers	X	❍			♦	♦♦		- well-drained loam	2-4 m (6½-13 ft.)
1	*Rosa woodsii*/ Prairie Rose	- from Prairies to B.C. - small scented flowers, bright red hips	X	❍			♦	♦♦		- well-drained	2 m (6½ ft.)
0b	Salix arctica / Arctic Willow	- alpine tundra - usually prostrate	X	❍			♦	♦♦	♦♦♦	- various	50 cm (1½ ft.)
2a	*Salix candida* / Hoary Willow	- from Yukon to Labrador - useful in masses		❍	◗			♦♦		- well-drained, alkaline soil	50 cm-3 m (1½-10 ft.)
6a	*Salix gracilistyla* / Rosegold Pussy Willow	- adaptable - graceful - ornamental catkins - bluish-grey leaves	X	❍	◗			♦♦		- various	2-5 m (6½-16 ft.)
3a	*Sambucus canadensis* / Common Elder	- from Lake Ontario to N.S. - white flowers, black berries that attract birds and are edible - suckers profusely—best used for naturalizing large sites	X, W	❍	◗		♦	♦♦	♦♦♦	- adaptable: sand, silt, loam, clay, prefers moist but will tolerate dry soil	3 m (10 ft.)
1	*Shepherdia canadensis* / Soapberry, Buffaloberry	- from Newfoundland to Alaska, except in some areas - tolerates poor soils, dry conditions, salt, easy to grow	X, W	❍			♦	♦♦		- sandy, rocky to loam, nutrient-poor soils	2 m (6½ ft.)
4	*Sorbus sitchensis* / Sitka Mountain Ash	- from coastal to eastern B.C. - red berries in fall		❍				♦♦	♦♦♦	- loamy soil	1-2 m (3-6½ ft.)
2a	*Symphoricarpos albus* / Snowberry	- from N.S. to B.C. - interesting white berries through winter - spreading	X, W	❍	◗	●	♦	♦♦		- tolerant of any soil	1.5 m (5 ft.)
2	*Vaccinium alaskaense* / Alaskan Blueberry	- from Alaska to Oregon - bluish-black berries with bluish flowers		❍			♦	♦♦		- heavy soils	1.5 m (5 ft.)
2	*Vaccinium myrtilloides* / Velvet leaved Blueberry	- from Labrador to B.C. and the Northwest Territories - berries commonly eaten, used for jams, pies, etc.		❍				♦♦	♦♦♦	- various: gravelly, sandy soil to clay loam - acidic, moist, cool soils	10-40 cm (4-16 in.)

Hardiness Zone	Latin name / Common name	Characteristics / Special visual features*	Useful for xeriscape (X), woodland (W) meadow/ prairie (P)	Sun exposure			Moisture			Soil	Height
				❍	◗	●	💧	💧💧	💧💧💧		
7b	*Vaccinium parvifolium* / Red Huckleberry	- from coastal B.C. - pinkish yellow flowers, bright red berries		❍	◗	●	💧	💧💧	💧💧💧	- rich, moist soil	1-4 m (3-13 ft.)
2a	*Viburnum lentago* / Nannyberry	- from south-east Saskatchewan to southern Quebec - white clusters of flowers - bluish-black fruit attracts birds	W	❍	◗	●		💧💧	💧💧💧	- all types of soils: sand, silt, loam, clay, organic - moist to wet	3-6 m (10-20 ft.)
2a	*Viburnum trilobum* / Highbush-cranberry	- from southern Manitoba to Atlantic - white clusters of flowers - bright red fruit attracts birds	W	❍	◗	●		💧💧	💧💧💧	- all types of soils: sand, silt, loam, clay, organic - moist to wet	4 m (13 ft.)
4	*Yucca filamentosa* / Adam's Needle	- broadleaf evergreen - long, pointed leaves with white flowers growing on a tall stalk	X	❍			💧	💧💧		- well-drained	60 cm (2 ft.)
PERENNIALS											
3	*Actaea rubra* / Red Baneberry	- from B.C., Prairie and eastern Canada - attractive berries that are poisonous	W		◗	●		💧💧	💧💧💧	- moist to average, nutrient-rich soil	70 cm (2 ft. 3 in.)
3	*Andropogon gerardii* / Big Bluestem	- from tall- and mixed-grass prairie regions of Canada - tall grass, bluish in summer, bronze in fall - drought tolerant	P, X	❍			💧	💧💧		- adaptable, moist to dry, clay to sandy, low fertility needs, tolerates heavy clay	90-240 cm (3-8 ft.)
4	*Anemone canadensis* / Canada Anemone	- from Prairie and eastern Canada - white flowers - spreading	P	❍	◗			💧💧		- adaptable (sand, silt, loam, clay) but best in moist, well-drained soil	40 cm (1 ft. 3 in.)
3	*Aquilegia canadensis* / Eastern Columbine	- from central and eastern Canada - lacy foliage, delicate flowers - self-seeds	W		◗	●		💧💧	💧💧💧	- adaptable, tolerates heavy clay soil, does well on moist, well-drained, loose soil	50 cm (1½ ft.)
3b	*Asclepias tuberosa* / Butterfly Weed	- from Prairie and eastern Canada - clusters of orange flowers attracts butterflies - drought tolerant	P, X	❍			💧	💧💧		- dry to average, well-drained	60-90 cm (2-3 ft.)
2	*Aster novae-angliae* / New England Aster	- from Prairie and eastern Canada - long-lasting purple flowers with yellow centres	X, P	❍	◗			💧💧		- adaptable, sand to heavy clay, moist to dry, but prefers moist	90-180 cm (3-6 ft.)
4	*Athyrium Filix-femina* / Lady Fern	- from Alaska to Atlantic - finely divided leaves	W		◗	●		💧💧	💧💧💧	- fertile, moist to wet, well-drained, neutral to acid soil	60 cm (2 ft.)
7b	*Blechnum spicant* / Deer Fern	- from B.C. - drought tolerant when in shade	W		◗	●		💧💧	💧💧💧	- rich, acidic, wet to avg.	30-60 cm (1-2 ft.)
2	*Campanula rotundifolia* / Common Harebell	- blue bell-shaped flowers	X	❍	◗	●	💧			- well-drained	30 cm (1 ft.)
3	*Coreposis lanceolata*/ Lance-leaved Coreopsis	- from Prairies and parts of Ontario - long-lasting yellow flowers - drought tolerant	P, X	❍			💧	💧💧		- well-drained dry to average soil, tolerates sandy infertile soil	30-60 cm (1-2 ft.)

3	*Dicentra formosa* / Western Bleeding Heart	- from B.C. - pink, heart shaped flowers	W		◗	●		♦♦	♦♦♦	- rich, moist, well-drained	45 cm (1½ ft.)
3	*Echinacea purpurea* / Purple Coneflower	- from Prairies - drought tolerant, easy to grow - large pink flowers	X,P	❍	◗		♦	♦♦		- well-drained	60-150 cm (2-5 ft.)
2a	*Elymus Canadensis* / Canada Wild Rye	- from Prairies - tall grass - drought tolerant	P	❍	◗			♦♦		- adaptable from clay to sand, tolerates heavy clay	90-150 cm (3-5 ft.)
3	*Eupatorium maculatum* / Joe-Pye Weed	- from Prairie and eastern Canada - tall, with pink flowers	P	❍	◗			♦♦	♦♦♦	- sand to clay, prefers moist soil	1.5 m-2 m (5-6½ ft.)
3	*Gentiana andrewsii* / Bottle Gentian	- from Prairie and eastern Canada - blue flowers, late summer	P	❍				♦♦	♦♦♦	- sand to clay, moist, rich to average soil	45 cm (1½ ft.)
3	*Helenium autumnale* / Sneezeweed	- from parts of Quebec to B.C. - daisy-like, yellow flowers (cultivars yellow to red), late summer	P	❍				♦♦		- sand to clay, average to moist soil	60-150 cm (2-5 ft.)
3	*Hepatica acutiloba* / Sharp-lobed Liverleaf	- from parts of Prairie and eastern Canada - good for dry woodland - white, pinkish flowers in early spring	W, X		◗	●	♦	♦♦	♦♦♦	- rich, moist to dry	10-15 cm (4-6 in.)
3	*Hosta spp.* / Plantain Lily, Hosta	- many cultivars and species with differing leaf colours, patterns and sizes - low, foliage plant suited to shady sites			◗	●		♦♦		- moist soil with good amount of organic matter	30-60 cm (1-2 ft.)
3	*Iris tenax* / Oregon Iris	- from B.C. - purple flowers - for moist sites see *Iris setosa*	P, X	❍			♦	♦♦		- well-drained, dry to average	40 cm (1 ft. 3 in.)
4b	*Lavandula angustifolia* / English Lavender	- fragrant leaves, purple flowers - herbaceous to semi-woody	X	❍	◗		♦	♦♦		- well-drained, does well in dry, poor soil (not heavy, wet soils)	40 cm (1 ft. 3 in.)
3	*Liatris aspera* / Rough Blazing Star	- from Prairie and eastern Canada - drought tolerant - spikes of pink-purple flowers	P, X	❍			♦	♦♦		- dry to average, neutral to acidic, well-drained, tolerates poor soil	60-120 cm (2-4 ft.)
2	*Lilium canadense* / Canada Lily	- from south-east Canada - striking yellow to red flowers	P	❍	◗			♦♦	♦♦♦	- moist to wet loam or clay	60-150 cm (2-5 ft.)
3	*Lupinus perennis* / Wild Lupine	- from Prairie and eastern Canada - drought tolerant - blue-violet flowers - legume: fixes nitrogen	X	❍	◗		♦	♦♦		- dry to average, well-drained	30-60 cm (1-2 ft.)
2	*Matteucia struthiopteris* / Ostrich Fern	- from Alaska to Newfoundland - easy to grow, spreading - edible fiddleheads	P	❍	◗	●			♦♦♦	- organic, moist to wet, slightly acid soil, sandy to clay loam	60-150 cm (2-5 ft.)
3	*Monardia didyma* / Bee Balm	- from eastern Canada - easy to grow - red flowers	P, X	❍	◗			♦♦	♦♦♦	- sand to clay	1 m (3 ft.)
2	*Monarda fistulosa* / Wild Bergamot	- from Prairies, eastern Canada - drought tolerant - pink flowers	P, X	❍			♦	♦♦		- adaptable, clay to sand, avg. to dry, fertile to nutrient poor	60-120 cm (2-4 ft.)
3	*Oenothera biennis* / Evening Primrose	- from Prairie and eastern Canada - drought tolerant, easy to grow - long lasting yellow flowers	X, P	❍			♦	♦♦		- clay, loam, sand, nutrient poor, dry to average	60 cm (2 ft.)

Hardiness Zone	Latin name / Common name	Characteristics / Special visual features*	Useful for xeriscape (X), woodland (W) meadow/ prairie (P)	Sun exposure			Moisture			Soil	Height
				❍	◗	●	♦	♦♦	♦♦♦		
3	*Osmunda cinnamomea* / Cinnamon Fern	- from Newfoundland to western Ontario - showy large fiddleheads	W	❍	◗	●		♦♦	♦♦♦	- moist, acidic, average to poorly drained soils	90-120 cm (3-4 ft.)
4	*Panicum virgatum* / Switchgrass	- from Prairie and eastern Canada - tall grass, yellow fall colour	P	❍				♦♦		- sand, silt, loam	70 cm - 2 m (2-6½ ft.)
3	*Phlox subulata* / Moss Phlox	- mounded groundcover with profuse white, pink or light blue flowers in early spring - moderate drought tolerance	X	❍	◗		♦	♦♦		- average, well-drained	5-10 cm (2-4 in.)
3	*Physostegia virginiana* / Obedient Plant	- from Prairie and eastern Canada - pink spike flowers - spreading	P	❍	◗			♦♦		- average, well-drained soil	1 m (3 ft. 3 in.)
3	*Polygonatum multiflorum* / Solomon's Seal	- from Prairie and eastern Canada - arching stems, lush green foliage, black berries in fall	W		◗	●		♦♦	♦♦♦	- sandy to clay, moist to average, prefers rich, cool, moist soil	70 cm (2 ft. 3 in.)
3	*Polystichum acrostichoides* / Christmas Fern	- from eastern Canada - evergreen, compact	W		◗	●		♦♦		- rich soil - prefers moist, but can tolerate dryness, acidic to neutral soil	30 cm - 60 cm (1-2 ft.)
3	*Rudbeckia hirta* / Black-eyed Susan	- from Prairie and eastern Canada - yellow daisy-like flowers with black centres, long-lasting - drought tolerant	P, X	❍			♦	♦♦		- sandy to clay soil, tolerates dry infertile soil	30-90 cm (1-3 ft.)
3	*Saponaria ocymoides* / Rock Soapwort	- profuse pink flowers - drought tolerant	P, X	❍			♦	♦♦		- grows well in poor, dry soil	25 cm (10 in.)
3	*Sedum spectabile* / Showy Stonecrop	- easy to grow - pink flower late summer, fall	X, P	❍			♦			- dry, well-drained	50 cm (1½ ft.)
3	*Smilacina racemosa* / False Solomon's Seal	- from across Canada - cluster of creamy-white flowers at end of stem - bright red berries - good for dry woodlands	W		◗	●		♦♦		- sandy to clay, prefers rich soil	80 cm (2½ ft.)
2	*Solidago Canadensis* / Canada Goldenrod	- from Prairie and eastern Canada - drought tolerant - try Zig Zag Goldenrod for shady, dry sites	P, X	❍			♦	♦♦	♦♦♦	- adaptable (fertility, texture, moisture), typically found on moist sites	1.10 m (3½ ft.)
4	*Sorghastrum nutans* / Indian Grass	- from Prairie and eastern Canada - drought tolerant - tall grass	P, X	❍			♦	♦♦		- average to dry, well-drained but tolerates heavy clay	90-240 cm (3-8 ft.)
3	*Tiarella cordifolia* / Foamflower	- from Prairie and eastern Canada - spikes of white flowers - good under trees	W		◗	●		♦♦	♦♦♦	- does best in rich, cool, moist soil	30 cm (1 ft.)
4	*Thymus vulgaris* / Thyme	- fragrant - low growing	X	❍			♦			- well-drained, dry	10 cm (4 in.)
4	*Verbena hastata* / Blue Vervain	- from Prairie and eastern Canada - spikes of purple flower	P	❍	◗			♦♦	♦♦♦	- sand to clay	90-150 cm (3-5 ft.)
4	*Viola canadensis* / Canada Violet	- from eastern Canada - white flowers with yellow centre	W		◗	●		♦♦		- prefers moist, rich soil, tolerates dry	20 cm (8 in.)

CLIMBING PLANTS												
3b	*Celastrus scandens* / American Bittersweet	- from Southern Manitoba and Ontario - best on large, poor sites - fruit attractive in arrangements			❍	◗			♦♦		- withstands about any soil condition, including dry	
3	*Clematis virginiana* / Virgin's Bower	- from Prairie and eastern Canada - aggressive, fast growing - white wispy flowers			❍	◗			♦♦		- adaptable, sand to clay	
	Lonicera dioica / Limber Honeysuckle	- from central Canada and southern Ontario	X, W		❍	◗		♦	♦♦		- moist to dry - gravelly, sandy to clayey loams	
3	*Parthenocissus quinquefolia* / Virginia Creeper	- from eastern Canada - fast growing, can be invasive - red fall colour, dark blue berries			❍	◗			♦♦		- tolerates about any kind of soil, difficult sites, city conditions	

**Eastern Canada includes Ontario and provinces to its east.*

For information on sources, refer to the "Resources and Further Reading" section at the end of this guide.

RESOURCES AND FURTHER READING

Barrier-free design, universal accessibility

Housing Choices for Canadians with Disabilities. Canada Mortgage and Housing Corporation; Ottawa, 1992.

Housing for Persons with Disabilities. Canada Mortgage and Housing Corporation; Ottawa, 1996.

Un logis bien pensé, j'y vis, j'y reste: Guide de rénovation pour rendre un logis accessible et adaptable. René Chamberland. Société d'habitation du Québec; Québec, 1999.

Bioengineering (soil)

Restoring Shorelines with Willows. Landowner Resource Centre and Ontario Ministry of Natural Resources, 1995. www.lrconline.com/Extension_Notes_English/pdf/willows.pdf

Soil Bioengineering: An Alternative to Concrete. Environment Canada, 2000. www.on.ec.gc.ca/doc/cuf_factsheets/soil-bioeng-e.html

Soil Bioengineering: Seminar Summary. Robbin Sotir. Harrington and Hoyle Ltd.; Markham, ON, 1984.

Soil Bioengineering and Ecological Systems Techniques. CD-ROM. Maccaferri Inc., 2001.

Sustainable Landscape Construction: A Guide to Building Outdoors. J. William Thompson and Kim Sorvig. Island Press; Washington, 2000.

Decks and deck materials

"ASTM Specifies Plastic Lumber for Exterior Decking" *ASTM Standardization News*, September 2001.

Builder's Guide to Decks. Leon A. Frechette. McGraw-Hill Publishing Company, 1996.

"Building a Healthy Deck" in *Healthy Housing Renovation Planner.* Canada Mortgage and Housing Corporation; Ottawa, 1999.

Decks: Step-by-step Techniques for Practical Design and Construction. Sunset Books; Menlo Park, CA, 1996.

Environmental Resource Guide. ed. Joseph A. Demkin. American Institute of Architects. John Wiley; New York, 1996.

Fact Sheet on Chromated Copper Arsenate (CCA) Treated Wood. Pest Management Regulatory Agency; Ottawa, 2003.

"Getting a Deck off to a Good Start". Peter J. Bilodeau. *Fine Homebuilding*, April/May 1999.

"A Green Wooden Deck". Ed Wyatt. *Home Energy*, January/Februrary 2001.

"Landscaping Projects: Detailing" and "Landscaping Projects: Exterior Decks". Canadian Wood Council; Ottawa. www.cwc.ca/design/landscaping_projects

Mohini Sain (Professor of Engineering, University of Toronto). Personal communication, 2003.

"On-deck: Composites" Peter Laks. *Smart Homeowner*, #5, May/June 2002.

Porches, Decks and Outbuildings. The Best of Fine Homebuilding. The Taunton Press, 1997

Selecting Wood for Outdoor Structures. Jeff Fahrenholtz. Sustainable Urban Landscape Information Series, University of Minnesota. www.sustland.umn.edu/implement/index.html

Sustainable Landscape Construction: A Guide to Green Building Outdoors. J. William Thompson and Kim Sorvig. Island Press; Washington, DC, 2000.

"Wood Treatment"; "Natural Durability of Major North American Softwoods"; "Durability of Wood Alternatives"; "Wood Durability/General"; and "Observations on Plastic Lumber as a Substitute for Preservative-treated Wood". *Wood Durability* website. Canadian Wood Council and Forintek Canada Corp. www.durable-wood.com/index.php

Green roofs and roof-top gardens

Introductory Manual for Greening Roofs. Cornelia Hahn Oberlander, Elisabeth Whitelaw and Eva Matsuzaki. Public Works and Government Services; Ottawa, 2002. ftp.pwgsc.gc.ca/rpstech/Service_Life_Asset_Management/PWGSC_GreeningRoofs_wLinks.pdf

Green Roofs for Healthy Cities website: www.greenroofs.ca/grhcc/index.html

"Merchandise Lofts Building Green Roof Case Study" and "Waterfall Building Green Roof Case Study". Canada Mortgage and Housing Corporation; Ottawa, 2002. www.cmhc-schl.gc.ca/en/imquaf/himu/buin_009.cfm

"NRC Green Roofs Evaluation Confirms Merit for Sustainable Building Design". Karen Liu. *National Housing Research Committee Newsletter*, Autumn 2003.

"On the Balcony". Sonia Day. *Gardening Life*, March/April 2002.

Hard surfaces, pavements and retaining walls

Environmental Resource Guide. ed. Joseph A. Demkin. American Institute of Architects. John Wiley; New York, 1996.

Guide pratique des travaux de votre aménagement paysager. B. Dumont, D. Lefebvre, M. Rousseau. Spécialités Terre-à-Terre; Québec, 1997.

Landscape Architectural Design and Maintenance. Canada Mortgage and Housing Corporation; Ottawa, 1982.

"Landscaping Projects: Detailing" and "Landscaping Projects: Retaining Walls". Canadian Wood Council; Ottawa. www.cwc.ca/design/landscaping_projects

L'aménagement des terrains en pente. Les publications du Québec, Gouvernement du Québec, 1993.

"Stone Walls in the Landscape". Gordon Hayward. *Landscape Trades*, June 2001.

Sustainable Landscape Construction: A Guide to Green Building Outdoors. J. William Thompson and Kim Sorvig. Island Press; Washington, DC, 2000.

Time-saver Standards for Landscape Architecture, 2nd edition. Charles W. Harris and Nicholas T. Dines. McGraw-Hill Publishing Company; New York, 1998.

Your Natural Home: The Complete Sourcebook and Design Manual for Creating a Healthy, Beautiful and Environmentally Sensitive Home. Janet Marinelli and Paul Bierman-Lytle. Little, Brown and Company; Toronto, 1995.

Walls, Walks and Patios. Creative Homeowner; New Jersey, 1997.

Hiring contractors

"Hiring a Contractor". *About Your House*, CE 26. Canada Mortgage and Housing Corporation; Ottawa, 2001. www.cmhc.ca/en/burema/gesein/abhose/abhose_ce30.cfm

Selecting a Landscape Contractor. Brochure from Landscape Ontario, Horticulture Trades Association.

Invasive plants/weeds

Integrated Pest Management -Weeds. Nova Scotia Agriculture and Fisheries. http://www.gov.ns.ca/nsaf/rir/weeds/index.htm

Invasive Plants. Canadian Botanical Conservation Network. www.rbg.ca/cbcn/en/invasives

Invasive Plants of Canada: Guide to Species and Methods of Control. Erich Haber. National Botanical Services, 2001. http://www.plantsincanada.com/

Invasive Plants of Natural Habitats in Canada: An Integrated Review of Wetland and Upland Species and Legislation Governing their Control. D.J. White, E. Haber and C. Keddy. Canadian Wildlife Service, 1993. www.cws-scf.ec.gc.ca/publications/inv/cont_e.cfm

Natural Invaders. Federation of Ontario Naturalists. www.ontarionature.org/enviroandcons/naturalinvaders/invasive.html

Ontario Weeds. Ontario Ministry of Agriculture and Food; Guelph, ON, 2003.

Ontario Weeds Gallery. Ontario Ministry of Agriculture and Food. http://www.gov.on.ca/OMAFRA/english/crops/facts/ontweeds/weedgal.htm

Weeds BC website: www.weedsbc.ca/

Weeds of Canada and the Northern United States. France Royer, Richard Dickinson. Lone Pine Publishing/The University of Alberta Press, 1999.

Landscape design, general landscape architecture

Designing the Natural Landscape. Richard L. Austin. Van Nostrand Reinhold; New York, 1984.

The Garden Design Book. Cheryl Merser. Harper Collins Publisher; New York, 1997.

Les grands principes de l'aménagement paysager. Denise Blais. Broquet; Ottawa, 1994.

Guide de construction en milieu naturel. Gouvernement du Québec, Direction générale des publications du Québec, 1984.

Graphic Standards for Landscape Architecture. R. Austin, T. Dunbar, K. Hulvershorn, W. Todd. Van Nostrand Reinhold; New York, 1986.

Handbook of Landscape Architectural Construction, vol 1 & 2. Maurice Nelischer. Landscape Architecture Foundation; Washington, 1985 and 1988.

Je dessine mon jardin. Daniel Puiboube. Éditions Marabout; Belgique, 1996.

Landscape Design: A Practical Approach, 2nd Edition. Leroy Hannebaum. Prentice Hall; Englewood Cliffs, N.J., 1990.

The Landscape Revolution. Andy Wasowski and Sally Wasowski. Contemporary Books, USA, 2000.

Second Nature: Adapting LA's Landscape for Sustainable Living. Patrick Condon and Stacy Moriarty, eds. TreePeople; Beverly Hills, CA, 1999.

The Yearbook of Landscape Architecture, Private Spaces in the Landscape. Richard L. Austin, Thomas R. Dunbar, Lane L. Marshall, Albert L. Rutledge, Frederick R. Steiner. Van Nostrand Reinhold; New York, 1998.

Land trusts and conservation easements

Conservation Easements. Nature Saskatchewan. www.unibase.com/~naturesk/easement.htm

Federation of Ontario Naturalists website. www.ontarionature.org/index.php3

Nature Conservancy of Canada website: www.natureconservancy.ca/files/index.asp

Lawns

Choosing Lawn Grasses. Department of Horticulture, Cornell University. www.hort.cornell.edu/gardening/lawn/grasses.html

Grass Seeding Tips. City of Toronto, 1998. www.city.toronto.on.ca/compost/seeding.htm

"Healthy Lawn Tips"; "Lawn Problems"; "Lawn Maintenance"; and "Lawn Ecology". Healthy Lawns fact sheets. Pest Management Regulatory Agency; Ottawa, 2002. www.healthylawns.net/english/html/hg-e.shtml

The Home Lawn. D.H. Taylor, D.B. White, W.C. Stienstra, M.E. Ascerno, Jr. University of Minnesota Extension Service, 1997. www.extension.umn.edu/distribution/horticulture/DG0488.html

Household Guide to Water Efficiency. Canada Mortgage and Housing Corporation; Ottawa, 2000.

"Lawn Care Through the Seasons"; "Managing Insects and Diseases on Home Lawns"; and "Managing Weeds on Home Lawns". Lawn fact sheets. Nova Scotia Agriculture and Fisheries, 1995. www.gov.ns.ca/nsaf/elibrary/archive/hort/lawn/

Low Input Lawn Care. Bob Mugaas. University of Minnesota Extension Service, 2000. Available on www.extension.umn.edu/distribution/horticulture/DG7552.html

Residential Landscapes: Comparison of Maintenance Time, Costs and Resources. Ecological Outlook. Canada Mortgage and Housing Corporation; Ottawa, 2000.

Low-maintenance Lawns. About Your House, Canada Mortgage and Housing Corporation; Ottawa, 2004.

The Natural Lawn and Alternatives. Brooklyn Botanic Garden; Brooklyn, 1995.

Turf Factsheets/Infosheets and Publications. Ontario Ministry of Agriculture and Food. www.gov.on.ca/OMAFRA/english/crops/hort/turf.html#factsheets

Turfgrass Species Selection. William E. Pound and John R. Street. Ohio State University Extension Fact Sheet. Ohio State University. http://ohioline.osu.edu/hyg-fact/4000/4011.html

Mosquitoes: Reducing breeding areas

Reduce Mosquito Breeding Areas Around Your Home. City of Toronto, Public Health. www.toronto.ca/health/pdf/wnv_breedingareas.pdf

West Nile Virus Surveillance Information. Health Canada. www.hc-sc.gc.ca/pphb-dgspsp/wnv-vwn/

Pests (also see Lawns)

The Backyard Bug Brigade. Environment Canada. www.ns.ec.gc.ca/epb/factsheets/bkyard_bug/bugs_brch.html

"Cross-country Pesticide Checkup: The debate continues". Rob Adams. *Landscape Trades*, April 2001.

Insect and Disease Problems. International Society of Arboriculture. Champaign, IL, 1995. www.treesaregood.com/treecare/insect_disease.asp

"What is Integrated Pest Management?" and "Preventing Pest Damage in Home Lawns". *Integrated Pest Management fact sheets.* Nova Scotia Environment and Labour. www.gov.ns.ca/enla/rmep/p2/ipm.htm

Integrated Pest Management Manual for Landscape Pests in British Columbia. L.A. Gilkeson and R.W. Adams. Ministry of Environment, Lands and Parks, 2000.

Pest Notes (fact sheets). Pest Management Regulatory Agency; Ottawa, 1997-2001. www.hc-sc.gc.ca/pmra-arla/english/consum/pnotes-e.html

Rodale's Successful Organic Gardening: Controlling Pests and Disease. Patricia S. Michalak and Linda Gilkeson. Rodale; Emmaus, PA, 1994.

Tree Diseases of Eastern Canada. D. T. Myren (Editor), Canadian Forestry Service Staff, G. Laflamme (Editor). Canadian Forestry Service, 1994.

Planting and plant maintenance (general)

Dirt Cheap Gardening: Hundreds of ways to save money in your garden. Rhonda Massingham Hart. Storey Communications, Inc.; Pownell, VT, 1995.

Easy Maintenance Gardening. Ken Burke and A. Cort Sinnes. Ortho Books, 1982.

The Garden That Cares for Itself. Greg Williams and Norm Rae. Ortho Books, 1990

Guide pratique des travaux de votre aménagement paysager. B. Dumont, D. Lefebvre, M. Rousseau. Spécialités Terre-à-Terre; Québec, 1997.

"Healthy trees and Shrubs" and "Roses and Other Flowers". *The Green Lane fact sheets.* Environment Canada, 1991. www.qc.ec.gc.ca/ecotrucs/solutionsvertes

At Home with Composting. The Composting Council of Canada. http://www.compost.org/backyard.html

"It's a Natural: Naturalize bulbs for minimum maintenance and maximum charm". Leslie Saffrey. *Landscape Trades*, September 2002.

Mulching Landscape Plants. Dr. Mary Ann Rose and Dr. Elton Smith. Ohio State University Extension Fact Sheet. Horticulture and Crop Science, Ohio State University; Columbus, OH. http://ohioline.osu.edu/hyg-fact/1000/1083.html

Mulches for Landscaping. Donald A. Rakow. Cornell University Department of Horticulture, 2000. www.hort.cornell.edu/gardening/fctsheet/mulch.html

Mulch Mania. Marie Hofer. Home and Garden Television. www.hgtv.com/hgtv/gl_soil_water_mulch/article

Norme - aménagement paysager à l'aide de végétaux (NQ 0605-100). Bureau de normalisation du Québec, première édition, 2001.

Residential Landscapes: Comparison of Maintenance Time, Costs and Resources. Ecological Outlook. Canada Mortgage and Housing Corporation; Ottawa, 2000.

Rodale's Illustrated Encyclopedia of Gardening and Landscaping Techniques. Barbara W. Ellis. Rodale Press; Emmaus, PA, 1990.

Plant Descriptions - Across Canada (also see "Sources used for Plant List" and "Woodland shade gardens, wildflower meadows/prairies")

Evergreen Native Plant Database. Evergreen. www.evergreen.ca/nativeplants/search/index.php

Home Grounds: Getting Started. Lorraine Johnson. Available on Evergreen website: www.evergreen.ca/en/hg/toolshed/getstarted/index.html (contains lists of field guides, native plant nurseries and plants for various regions of Canada)

Native Trees of Canada. R.C. Hosie. Fitzhenry & Whiteside Limited; Markham, Ontario, 1990.

On the Living Edge: Your Handbook for Waterfront Living (four editions—BC/Yukon, Alberta, Saskatchewan/Manitoba, Ontario). Sarah Kipp and Clive Callaway. Federation of British Columbia Naturalists, Rideau Valley Conservation Authority, 2002-2003.

Plant Hardiness Zones of Canada, 2000. Agriculture Canada and Agri-Food Canada, Natural Resources Canada, and National Atlas of Canada. http://sis.agr.gc.ca/cansis/nsdb/climate/hardiness/ or http://atlas.gc.ca/site/english/maps/forest/forestcanada/planthardi

Trees in Canada. John Laird Farrar. Fitzhenry and Whiteside Limited; Markham, Ontario, 1995.

Plant Descriptions - British Columbia

Indicator Plants of Coastal British Columbia. Karel Klinka, V.J. Krajina K. Klinka, A. Ceska and A.M. Scagel. UBC Press; Vancouver, 1989.

Gardening with Native Plants of the Pacific Northwest. Arthur R. Kruckeberg. University of Washington Press; Seattle, 2003.

Native Plants in the Coastal Garden: A Guide for Gardeners in British Columbia and the Pacific Northwest. April Pettinger. Whitecap Books; North Vancouver, 1996.

Naturescape British Columbia: Caring for Wildlife Habitat at Home, Native Plant and Animal Booklet. Susan Campbell. Naturescape British Columbia; Victoria, 1995.

Plants of Coastal British Columbia. Jim Pojar and Andy MacKinnon. Lone Pine Publishing; Vancouver, 1994.

Plants of Southern Interior British Columbia. Robert Parish and Ray Coupé and Dennis Lloyd, Lone Pine Publishing; Vancouver, 1996.

Trees, Shrubs and Flowers to Know in British Columbia and Washington. C.P. Lyons and Bill Merilees. Lone Pine Publishing; Vancouver, 1995.

Trees and Shrubs of British Columbia. Brayshaw, Christopher, 1996. University of British Columbia Press, Vancouver, Canada.

Plant Descriptions - Prairie Canada

Conservation of Canadian Prairie Grasslands - A Landowner's Guidebook. Gary C. Trottier. Environment Canada, Canadian Wildlife Service, 2002.

Manitoba's Tall Grass Prairie: A Field Guide to an Endangered Space. T. Reaume. Manitoba Naturalists Society; Winnipeg, 1993.

Manitoba Wayside Flowers. Linda Kershaw. Lone Pine Publishing; Edmonton, 2002.

The Plants of the Western Boreal Forest and Aspen Parkland. Derek Johnson, Jim Pojar, Andy MacKinnon and Linda Kershaw. Lone Pine Publishing; Edmonton, 1995.

Saskatchewan Wayside Flowers. Linda Kershaw. Lone Pine Publishing; Edmonton, 2002.

Trees and Shrubs of Alberta. Kathleen Wilkinson. Lone Pine Publishing; Edmonton, 1995.

Wildflowers Across the Prairies. J. R. Jowsey. F.R., Vance, J.S. McLean and F.A Switzer. Greystone Books; Toronto, 1999.

The Wildflower Gardener's Guide: Great Plains and Canadian Prairies Edition. Henry W. Art. Garden Way Publishing; Pownal, VT, 1991.

Plant Descriptions - Ontario, Quebec, Atlantic

"Plantes du Québec". *Bioclic.ca.* http://www.bioclic.ca/encyclopedie/plantes/plantes.html

"Forest Flora" and "Prairie and Savana Flora". *Carolinian Canada* website. www.carolinian.org/SpeciesHabitats_RepSpecies.htm

A Field Guide to Wildflowers: Northeastern and North-Central North America. Roger Tory Peterson and Margaret McKenny. Houghton Mifflin Co., 1998.

A Field Guide to Eastern Trees. Janet Wehr and George A. Petrides. Houghton Mifflin Co., 1998.

La Flore du Québec. Site Internet. http://www.floreduquebec.cjb.net/

Flowers of the Wild: Ontario and the Great Lakes Region. Zile Zichmanis and James Hodgins. Oxford University Press; Toronto, 1982.

Forest Plants of Central Ontario. Brenda Chambers, Karen Legasy and Cathy V. Bentley. Lone Pine Publishing, 1996.

Forest Plants of Northeastern Ontario. Karen Legasy, Shayna LaBelle-Beadman, & Brenda Chambers. Lone Pine Publishing, 1995.

Ontario Wildflowers. Linda Kershaw. Lone Pine Publishing, 2002.

Plants of Carolinian Canada. Larry Lamb and Gail Rhynard. Federation of Ontario Naturalists; Don Mills, ON, 1994.

QuebecPaysage.com: La référence du paysage Québécois. www.quebecpaysage.com/ResultatsRecherche.aspx

Restoring Nature's Place: A Guide to Naturalizing Ontario Parks and Greenspace. Jean-Marc Daigle and Donna Havinga. Ecological Outlook Consulting and Ontario Parks Association, 1996.

Shrubs of Ontario. James H. Soper and Margaret L. Heimburger. Royal Ontario Museum; Toronto, 1982.

Liste des végétaux indigènes. La Société de l'arbre du Québec. www.sodaq.qc.ca/pepinieres/produits.asp?liste=0&type=1

Trees of the Carolinian Forest: A Guide to Species, Their Ecology and Uses. Gerry Waldron. Boston Mills Press Inc., 2003.

Trees of Knowledge: A Handbook of Maritime Trees. Geoffrey A. Ritchie. Tay Tree & Land Services, Canadian Forest Service, Natural Resources Canada, Atlantic Forest Centre, 1996. www.atl.cfs.nrcan.gc.ca/index-e/what-e/publications-e/afcpublications-e/maritimetrees-e/maritimetrees-e.html

Trees of Nova Scotia. Gary Saunders. Nimbus Publishing and Nova Scotia Department of Natural Resources, 1995.

Trees of Ontario. Linda Kershaw. Lone Pine Publishing, 2001.

Wetland Plants of Ontario. Alan G. Harris, Linda Kershaw and Steven G. Newmaster. Lone Pine Publishing, 1997.

The Wildflower Gardener's Guide: Northeast, Mid-Atlantic, Great Lakes and Eastern Canada Edition. Henry W. Art. Garden Way Publishing; Pownal, VT, 1991.

Plant Descriptions - websites (searchable plant databases)

The PLANTS Database, Version 3.5 (http://plants.usda.gov). USDA, NRCS, 2002. National Plant Data Center, Baton Rouge, LA 70874-4490 USA.

Ornamental Plants plus Version 3.0. Michigan State University Extension and the Michigan Nursery and Landscape. http://www.msue.msu.edu/msue/imp/modzz/masterzz.html.

University of Connecticut Plant Database. http://www.hort.uconn.cdu/Plants/index.html

Shorelines, living by water bodies (creeks, lakes, oceans, etc.) and natural areas

The Living by Water Project website. www.livingbywater.ca

On the Living Edge: Your Handbook for Waterfront Living (four editions—BC/Yukon, Alberta, Saskatchewan/Manitoba, Ontario). Sarah Kipp and Clive Callaway. Federation of British Columbia Naturalists, Rideau Valley Conservation Authority, 2002-2003.

Neighbours of Mississauga's Natural Areas. A Comprehensive Information Booklet for Property Owners Living Near Natural Areas. City of Mississauga, 2001.

Preserving and Restoring Natural Shorelines. Landowner Resource Centre and Ontario Ministry of Natural Resources, 2000. www.lrconline.com/Extension_Notes_English/pdf/shrlns.pdf

The Shore Primer: A Cottager's Guide to Healthy Waterfront. Ray Ford. Cottage Life in Association with Fisheries and Oceans Canada; Ottawa, 1999. www.dfo-mpo.gc.ca/canwaters-eauxcan/infocentre/guidelines-conseils/guides/shore-primer/fpd/fullprintable_fpd_e.asp

Soil analysis and grading

Accredited Soil Testing Labs. Ontario Ministry of Agriculture and Food. http://www.gov.on.ca/OMAFRA/english/crops/resource/soillabs.htm

Dirt Cheap Gardening: Hundreds of ways to save money in your garden. Rhonda Massingham Hart. Storey Communications, Inc. Pownel, VT, 1995.

Field Manual for Describing Soils in Ontario. K.A. Denholm and L.W. Schut. Ontario Centre for Soil Resource Evaluation, Guelph Agricultural Centre; Guelph, ON, 1993.

Household Guide to Water Efficiency. Canada Mortgage and Housing Corporation; Ottawa, 2000.

"Interaction Between Trees, Sensitive Clay Soils and Your Foundation". *About Your House CE 31.* Canada Mortgage and Housing Corporation; Ottawa, 2002. http://www.cmhc.ca/en/burema/gesein/abhose/abhose_ce45.cfm

Norme - aménagement paysager à l'aide de végétaux (NQ 0605-100). Bureau de normalisation du Québec, première édition, 2001.

Restoring Nature's Place: A Guide to Naturalizing Ontario Parks and Greenspace. Jean-Marc Daigle and Donna Havinga. Ecological Outlook Consulting and Ontario Parks Association, 1996.

Rodale's Illustrated Encyclopedia of Gardening and Landscaping Techniques. Barbara W. Ellis. Rodale Press; Emmaus, PA, 1990.

Rodale's Successful Organic Gardening: Improving the Soil. Erin Hynes. Rodale Press; Emmaus, PA, 1994.

The Secret to Successful Living - Healthy Soil. The Composting Council of Canada. http://www.compost.org/healthysoilPR.html

Sustainable Landscape Construction: A Guide to Green Building Outdoors. J. William Thompson and Kim Sorvig. Island Press; Washington DC, 2000.

Sun and wind control

Designing and Caring for Windbreaks. Landowner Resource Centre and Ontario Ministry of Natural Resources, 1995. www.lrconline.com/Extension_Notes_English/forestry/wndbrk.html

Energy-Efficient and Environmental Landscaping. Ann Simon Moffat and Marc Schiler. Appropriate Solutions Press; South Newfane, VT, 1993.

"Energy Savings with Trees". Gordon M. Heisler. *Journal of Arboriculture*, May 1986.

"Plantings that Save Energy". Gordon M. Heisler and David R. DeWalle. American Forests, September 1984.

Tap the Sun: Passive Solar Techniques and Home Designs. Canada Mortgage and Housing Corporation; Ottawa, 1998.

Stormwater and rain gardens

Canadian Wood-frame House Construction. Canada Mortgage and Housing Corporation; Ottawa, 1997.

"Grassed Swales" in *Evaluation of Management of Highway Runoff Water Quality.* G. Kenneth Young, Stuart Stein, Pamela Cole, Traci Kammer, Frank Graziano, Fred Bank. Federal Highway Administration, U.S. Dept. of Transportation, 1996.

Investigating Diagnosing and Treating Your Damp Basement. Canada Mortgage and Housing Corporation; Ottawa, 1992.

Minnesota Urban Small Sites BMP Manual. Barr Engineering Company. Metropolitan Council Environmental Services; St. Paul, Minnesota, 2001. http://www.metrocouncil.org/environment/Watershed/BMP/manual.htm

"Plotting to Infiltrate? Try Rain Gardens". Lorrie Stromme. *Yard and Garden News,* May 2001. Available at the University of Minnesota Extension Service website: www.extension.umn.edu/yardandgarden/YGLNews/YGLN-May0101.html#rain

Rain Gardens: A how-to manual for homeowners. Roger Bannerman and Ellen Considine. University of Wisconsin Extension: Madison, WI, 2003. http://clean-water.uwex.edu/pubs/raingarden/rgmanual.pdf

"Rain Gardens". Karen Cozetto. *Conscious Choice,* May 2001. www.consciouschoice.com/environs/raingardens1405.html

"Site Design Solutions for Achieving Performance Targets" in *Stormwater Planning: A Guidebook for British Columbia.* Kim A. Stephens, Patrick Graham, David Reid. B.C. Ministry of Water, Land and Air Protection, 2002. http://wlapwww.gov.bc.ca/epd/epdpa/mpp/stormwater/chapter7.pdf

Stormwater Management Practices Planning and Design Manual. Ontario Ministry of Environment; Toronto, 1994, updated in 2003. http://www.ene.gov.on.ca/envision/gp/4329eindex.htm

Stormwater Pollution: Tips on how you can improve water quality. Works and Emergency Services, City of Toronto, 2000.

Tree benefits

Urban Forest Values: Economic Benefits of Trees in Cities. Center for Urban Horticulture, University of Washington, 1998. http://www.cfr.washington.edu/Research/fact_sheets/29-UrbEconBen.pdf

"Value of Urban Trees" in Technical Guide to Urban and Community Forestry. USDA Forest Service. http://www.na.fs.fed.us/spfo/pubs/uf/techguide/toc.htm

Tree planting and care

Arboriculture: Care of Trees, Shrubs and Vines in the Landscape. Richard W. Harris. Prentice-Hall Inc.; Englewood Cliffs, NJ, 1983

Common Pests of Trees in Ontario. Ontario Ministry of Natural Resources; Toronto, 1995.

Field Guide to Tree Diseases in Ontario. NRCan; Sault Ste. Marie, Ontario, 1995.

Guide pratique des travaux de votre aménagement paysager. B. Dumont, D. Lefebvre, M. Rousseau. Spécialités Terre-à-Terre; Québec, 1997.

A Guide to Tree Planting. Canadian Forest Service, Natural Resources Canada; Ottawa, 1992. Available on Tree Canada Foundation website: http://www.treecanada.ca/publications/guide.htm

"Helping Your Trees Survive Storm Damage". *About Your House, CE12.* Canada Mortgage and Housing Corporation; Ottawa, 1999.

How to Prune Trees. Peter J. Bedker, Joseph G. O'Brien and Manfred E. Mielke. USDA Forest Service. http://www.na.fs.fed.us/spfo/pubs/howtos/ht_prune/prun001.htm.

"Mature Tree Care"; "Insect and Disease Problems"; "New Tree Planting"; "Pruning Young Trees"; and "Pruning Mature Trees". *Tree Are Good* website. International Society of Arboriculture. http://www.isa-arbor.com/consumer/consumer.html

Maintaining Healthy Urban Trees. Landowner Resource Centre and Ontario Ministry of Natural Resources, 2000. www.lrconline.com/Extension_Notes_English/pdf/urbntrs.pdf

A New Tree Biology: Facts, Photos and Philosophies on Trees and their Problems and Proper Care. Alex L. Shigo. Sherwood Dodge Printers; Littleton, NH, 1993.

Nine Things You Should Know About Trees. The National Arbour Day Foundation. http://www.arborday.org/trees/nineThings.html

Norme - aménagement paysager à l'aide de végétaux (NQ 0605-100). Bureau de normalisation du Québec, première édition, 2001.

Tree preservation

Avoiding Tree Damage During Construction. International Society of Arboriculture; Champaign, IL, 1995.

"Saving Trees During Construction". Paul Fisette and Dennis Ryan. *Journal of Light Construction*, September 2001.

Protecting Existing Landscape Trees from Construction Damage Due to Grade Changes. Everett Janne and Douglas F. Welsh. Texas Agricultural Extension Service, (L 1309), 1978. http://aggie-horticulture.tamu.edu/extension/ornamentals/protect/protect.html

Sustainable Landscape Construction: A Guide to Green Building Outdoors. J. William Thompson and Kim Sorvig. Island Press; Washington, DC, 2000.

Water gardens/ponds

"Les cascades et les jardins d'eau" *Fleurs, plantes et jardins*, volume 2, number 2, May 1991.

Complete Guide to Water Gardens. Kathleen Fisher. Creative Homeowner; New Jersey, 2000.

Dreamscapes Magalog. OASE Dreamscapes; Ventura, CA. Brochure.

Guide pratique des travaux de votre aménagement paysager. B. Dumont, D. Lefebvre, M. Rousseau. Spécialités Terre-à-Terre; Québec, 1997.

Planting the Seed: A Guide to Establishing Aquatic Plants. Andy Hagen. Environment Canada, Environmental Conservation Branch, 1996.

Water Gardening. Sheridan Nurseries. www.sheridannurseries.com/GardenSite/sectiongardeninginformationFRAMESET.htm

"West Nile Water Gardens". Carol Matthews. *Landscape Trades*, June 2003.

Wildlife gardens (also see Woodland shade gardens, wildflower meadows/prairies)

Attracting Animal Life to your Garden. Larry Lamb. Available on The Caretaker web site, http://wmuma.com/caretaker/naturalization/llamb.html

"Butterfly Gardens Can Be Elegant". Theresa M. Forte. Landscape Trades, April 2001.

Creating Habitat for Wildlife - A Garden Planner. The Canadian Wildlife Federation; Ottawa, 1996.

Natural Gardening. Jim Knopf, Sally Wasowski, John Kadel Boring, Glenn Keater, Jane Scott, Erica Glasener. The Nature Company; Berkeley, 1995.

NatureScape Alberta: Creating and Caring for Wildlife Habitat at Home. Myrna Pearman and Ted Pike. Red Deer River Naturalists and Federation of Alberta Naturalists; Red Deer, 2000.

Naturescape BC website. www.hctf.ca/naturescape/resources.htm (includes the journal Naturescape News: A Forum for Naturescapers)

Naturescape British Columbia: Caring for Wildlife Habitat at Home, Provincial Guide. Susan Campbell and Sylvia Pincott. Naturescape British Columbia; Victoria, 1995.

Woodland shade gardens, wildflower meadows/prairies

100 Easy to Grow Native Plants: For North American Gardens in Temperate Zones. Lorraine Johnson. Fire Fly Books; Toronto, 1999.

City Form and Natural Process. Michael Hough. Routledge; New York, 1984.

"Mulches Help Trees Beat Weed Competition"; "Tree Guards Protect Your Trees"; and "Cover Crops Help Tree Seedlings Beat Weed Competition". *Extension Notes.* Landowner Resource Centre and Ontario Ministry of Natural Resources, 1994-1999.
www.lrconline.com/Extension_Notes_English/forestry/for_index.html

A Garden of Wildflowers: 101 Native Species and How to Grow Them. Henry W. Art. Storey Communications, Inc.; Pownal, VT, 1986.

Grow Wild! Native Plant Gardening in Canada and Northern United States. Lorraine Johnson. Random House of Canada; Toronto, 1999.

A Guide to Natural Woodland and Prairie Gardening. Robert Dorney and Jo Rich. Natural Woodland Nursery. Ecoplans, Ltd.; Waterloo, 1978.

Home Grounds: Getting Started. Lorraine Johnson. Available on Evergreen website: www.evergreen.ca/en/hg/toolshed/getstarted/index.html

How to Naturally Landscape Without Aggravating Neighbors and Village Officials. Bret Rappaport. Wild Ones Handbook: Wild Ones -Natural Landscapers, Ltd. and Great Lakes National Program Office. Available on EPA website: www.epa.gov/greenacres/wildones/handbk/wo10.html

Natural Landscaping: Designing with Native Plant Communities. John Dieklemann and Robert Schuster. McGraw-Hill; New York, 1982. Second edition, 2003.

Naturalizing Your City Backyard. Walter Muma. The Caretaker website. http://wmuma.com/caretaker/naturalization/llambnatyard.html

The Ontario Naturalized Garden. Lorraine Johnson. Whitecap; Vancouver, 1995.

Planting the Seed: A Guide to Establishing Prairie and Meadow Communities in Southern Ontario. Kim Delaney, Rodger P. Lindsay, Allen Woodliffe, Gail Rhynard and Paul Morris. Environment Canada, Environmental Conservation Branch, 2000.

Residential Landscapes: Comparison of Maintenance Time, Costs and Resources. Ecological Outlook. Canada Mortgage and Housing Corporation; Ottawa, 2000.

Restoring Nature's Place: A Guide to Naturalizing Ontario Parks and Greenspaces. Jean-Marc Daigle and Donna Havinga. Ecological Outlook Consulting and Ontario Parks Association, 1996.

Xeriscapes and water conservation

Creating the Prairie Xeriscape: Low Maintenance, Water-efficient Gardening. Sara Williams. University Extension Press, University of Saskatchewan; Saskatoon, 1997.

Dry-land Gardening: A Xeriscaping Guide for Dry-summer, Cold-winter Climates. Jennifer Bennet. Firefly Books Ltd.; Willowdale, ON, 1998.

Household Guide to Water Efficiency. Canada Mortgage and Housing Corporation; Ottawa, 2000.

Lawn Watering and Soil Type. City of Toronto, 1998.
http://www.city.toronto.on.ca/watereff/watering.htm

Xeriscaping: Water-Conserving Landscaping. City of Toronto, 1998.
http://www.city.toronto.on.ca/compost/xeriscap.htm

Xeriscape Principles. Denver Water Conservation.
http://www.denverwater.org/cons_xeriscape/cons_xeriscapeframe.html

Sources used for plant list

100 Easy-to-Grow Native Plants for Canadian Gardens. Lorraine Johnson. Random House of Canada; Toronto, 1999.

Botanix Guide, 7th Edition. Rona inc.; Boucherville, Quebec, 1999.

Grow Wild!: Native Plant Gardening in Canada. Lorraine Johnson. Random House of Canada; Toronto, 1999.

Manual of Woody Landscape Plants. Michael A. Dirr. Stripes Publishing Company; Champaign, Illinois, 1983.

Native Trees of Canada. R.C. Hosie. Fitzhenry & Whiteside Limited; Markham, ON, 1990.

Plantes sauvages des lacs, rivières et tourbières. Fleurbec (éditeur); Saint-Augustin (Portneuf), Québec, 1987.

Plantes sauvages des villes et des champs. Fleurbec (éditeur); Saint-Augustin (Portneuf), Québec, 1978.

Plantes sauvages printanières. Gisèle Lamoureux et collaborateurs. Fleurbec (éditeur); Saint-Augustin (Portneuf), Québec, 1975.

Plantes sauvages des villes, des champs et bordure des chemins. Fleurbec (éditeur); Saint-Augustin (Portneuf), Québec, 1983.

Répertoire des arbres et arbustes ornementaux. Hydro-Québec, 1998.

Restoring Nature's Place: A Guide to Naturalizing Ontario Parks and Greenspaces. Jean-Marc Daigle and Donna Havinga. Ecological Outlook Consulting and Ontario Parks Association, 1996.

Trees in Canada. John Laird Farrar. Fitzhenry and Whiteside Limited; Markham, Ontario, 1995.

David Tarrant (UBC Botanical Garden and Centre for Horticulture). Personal communication, 2003.

Scott Martin (Wild Canada, Wasaga Beach, Ontario) Personal communication, 2003.

Cornelia Haun Oberlander, Landscape Architect, Vancouver. Written input, 2002.

The PLANTS Database, Version 3.5 (http://plants.usda.gov). USDA, NRCS, 2002. National Plant Data Center, Baton Rouge, LA 70874-4490 USA.

Ornamental Plants plus Version 3.0. Michigan State University Extension and the Michigan Nursery and Landscape. http://www.msue.msu.edu/msue/imp/modzz/masterzz.html.

University of Connecticut Plant Database. http://www.hort.uconn.edu/Plants/index.html

Associations of Landscape Architects:

Alberta	www.aala.ab.ca
Atlantic	www.apala.net
British Columbia	www.bsla.org
Manitoba	www.mala.net
Ontario	www.oala.on.ca
Quebec	www.aapq.org
Saskatchewan	www.sala.sk.ca

Organizations and Associations:

Alberta Native Plant Council	www.anpc.ab.ca
BC Landscape and Nursery Association	www.gardenwise.bc.ca
Canadian Botanical Conservation Network	www.rbg.ca/cbcn
Canadian Nature Federation	www.cnf.ca
Canadian Nursery Landscape Association	www.canadanursery.com
Canadian Wildlife Foundation	www.wildaboutgardening.org
Carolinian Canada	www.carolinian.org
Evergreen www.evergreen.ca	
Federation of Ontario Naturalists	www.ontarionature.org
Go for Green	www.goforgreen.ca
Green Roofs for Healthy Cities	www.greenroofs.ca/grhcc/index.html
International Society of Arboriculture	www.isa-arbor.com
La société de l'arbre du Quebec	sodaq@cfl.forestry.ca
Landowner Resource Centre	lrconline.com
Landscape Ontario	http://www.hort-trades.com/
Living by Water Project	www.livingbywater.ca

Living Prairie Museum	www.city.winnipeg.mb.ca/cms-prod/parks/envserv/interp/living.html
Native Plant Society of Saskatchewan	www.npss.sk.ca
NatureScape Alberta	http://www.naturescape.ab.ca
Naturescape British Columbia	http://www.hctf.ca/nature.htm
North American Native Plant Society	www.nanps.org
Ontario Native Plants	www.nativeplants.on.ca
QuebecPaysage.com: La référence du paysage Québécois	www.quebecpaysage.com/ResultatsRecherche.aspx
Royal Botanical Gardens	www.rbg.ca/
Tallgrass Ontario	www.tallgrassontario.org
Tree Canada Foundation - Tree Plan Canada	www.treecanada.ca/
Wildflower Society of Newfoundland and Labrador	http://www.chem.mun.ca/~hclase/wf/index.htm